Third-Generation and Wideband HF Radio Communications

For a listing of recent titles in the
Artech House Mobile Communications Series,
turn to the back of this book.

Third-Generation and Wideband HF Radio Communications

Eric E. Johnson
Eric Koski
William N. Furman
Mark Jorgenson
John Nieto

ARTECH HOUSE
BOSTON | LONDON
artechhouse.com

Library of Congress Cataloging-in-Publication Data
A catalog record for this book is available from the U.S. Library of Congress.

British Library Cataloguing in Publication Data
A catalog record for this book is available from the British Library.

ISBN-13: 978-1-60807-503-4

Cover design by Vicki Kane

685 Canton Street
Norwood MA 02062

10 9 8 7 6 5 4 3 2 1

Contents

Preface

High-frequency (HF) radio technology has made tremendous strides since the publication of *Advanced High-Frequency Radio Communications* by Johnson, Desourdis, Earle, Cook, and Ostergaard in 1997. Even as that book was going to press, the third generation of automatic link establishment (ALE) technology was well along in its development, as were data modem waveforms that offered a fourfold increase in data rates. In the first decade of the twenty-first century, the demand for ever-higher data throughputs impelled the HF technology and regulatory communities to consider breaking out of the long-standing 3-kHz channelization of the HF spectrum. As a result of several years' work on wideband HF technology, we now have new standards in the United States for radios and waveforms capable of fully utilizing channels of up to 24 kHz: MIL-STD-188-141C and MIL-STD-188-110C. With these two significant advances in HF radio technology—third-generation (3G) technology and wideband HF (WBHF) waveforms–since *Advanced High-Frequency Radio Communications* was published, it was clearly time for a new book that documents both the new capabilities and the thinking that went into them.

This is a book about technology, a celebration of innovation in HF radio at the dawn of the twenty-first century. We describe the technology, offer ideas for applying it, and evaluate the performance of 3G and wideband HF systems using both simulation and real-world measurements. Technical details are included throughout the book, but they may be skipped by readers interested in applying the new technologies to practical problems rather than in designing equipment.

The first four chapters briefly summarize HF radio communications. We cover topics needed later in the book, and refer the reader to the existing literature for additional depth. In these chapters, we review twentieth century HF radio technology, including ALE and narrowband data waveforms and protocols. Chapter 1 presents historical perspective and an introduction to HF radio. Chapter 2 describes the challenging ionospheric channel that is used for communicating beyond line-of-sight. In Chapter 3, we discuss the state-of-the-art serial-tone modem waveforms and protocols that were still in development when *Advanced High-Frequency Radio Communications* was in preparation, and consequently, were not documented in that book. Chapter 4 summarizes the second-generation ALE system (well-documented in the previous book), so that we have a baseline for comparison with 3G ALE.

Chapter 5 is devoted to a thorough discussion of 3G technologies, an integrated suite that includes ALE, data transfer, traffic management, and automatic link maintenance protocols. The scalable burst modem technology that underlies the 3G protocols is a key contributor to the superior performance of 3G systems: by lowering the signal-to-noise ratio required to communicate, these burst waveforms permit 3G radios to operate at reduced transmitter power levels compared to 2G systems. Lower operating power reduces both channel congestion and battery power consumption in the tactical networks where 3G has seen wide use. The 3G suite was also optimized for synchronous operation, which provides both faster linking and, again, reduced channel congestion compared to 2G systems.

Wideband HF data waveforms are the subject of Chapter 6. This technology was standardized in the United States in September 2011, as the writing of this book commenced. Implementation and testing of the technology had progressed in parallel with standardization, so we are able to include here not only a description and discussion of the WBHF technologies, but also measurements collected during over-the-air demonstrations of WBHF radios and modems. Among the exciting new capabilities achieved by WBHF is communicating real-time, full-motion video over thousands of kilometers via HF skywave channels.

Finally, in Chapter 7, we look ahead to possible next steps in the advancement of HF radio technology, including the possible application of cognitive radio techniques to the problem of finding usable portions of wideband HF channels in the presence of dynamic interference.

We hope that this book will be useful as a reference for HF system architects and engineers, and that it will inspire the research community to continue to advance the science and technology of HF radio communications.

CHAPTER 1

HF Radio

The history of radio now goes back over 100 years. At the start of the twentieth century, Marconi and other radio pioneers[1] were building large spark gap transmitters that used long wavelengths to communicate with ships at sea and even to compete with trans-Atlantic cables. In the pursuit of greater range, the size, operating wavelength, and input power of these commercial radiotelegraph transmitters were continually increased, eventually reaching hundreds of kilowatts and wavelengths of thousands of meters [2].

Shorter wavelengths (200 m or less) seemed, at the time, to have little commercial value and were granted to radio amateurs for experimental use [3, 4]. Thus it fell to the "hams" to discover the happy confluence of features found in the shortwave or high frequency (HF) band:

- Global range is achievable with far less power than was employed by the commercial radiotelegraph services. This higher efficiency is due to the ionosphere (see Chapter 2), whose existence was only hypothetical in the first decade of the twentieth century.
- Atmospheric noise is lower in the HF band than at the longer wavelengths used by Marconi.
- Antennas are much easier to build for wavelengths of tens of meters than for kilometer wavelengths.

Thus, whereas the commercial radiotelegraph operations required hundreds of kilowatts and monumental antenna structures for a longwave trans-Atlantic circuit, radio amateurs today communicate across the Atlantic with a backyard antenna and a few watts by using the HF band.

HF radio is now widely used, not only by the amateur community, but also by governmental and nongovernmental agencies worldwide whenever an alternative (or backup) to satellites for over-the-horizon communications is needed (see Figure 1.1):

- Ships at sea;
- Aircraft out of range of line-of-sight radio networks;
- Military operations;

1. For example, Telefunken of Germany, United Wireless of the United States, and the Lodge-Muirhead Syndicate of the United Kingdom were also in the radiotelegraph business [1].

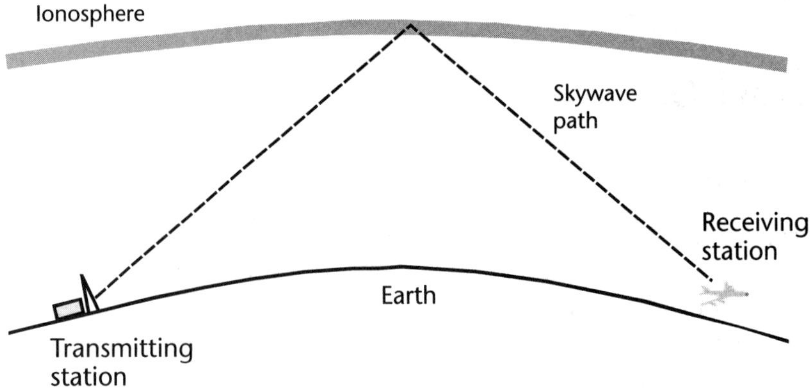

Figure 1.1 HF radio.

- Disaster areas where the terrestrial communications infrastructure has been destroyed or is overburdened;
- Distant regions lacking other communications.

Formally, the HF band denotes frequencies between 3 and 30 MHz, although HF technology is also used in the upper parts of the medium frequency (MF) band, down to perhaps 2 MHz.

1.1 HF Radio Transmissions

The early spark-gap transmitters evolved from Hertz's original experimental apparatus in which stored energy was discharged across a spark gap within a tuned circuit. The resulting damped oscillation, coupled into radiating conductors, produced pulses of radio waves that could be detected at a distance with a suitable resonant circuit.

This crude technology sufficed when radio was in its infancy and few transmitters were in operation. However, as the uses for radio grew rapidly in the early twentieth century, the need for narrowband radios became apparent: the "dirty" transmissions from spark gap transmitters readily interfered with the reception of other wireless signals, either by accident [5] or by intent [6].

The progression from spark technology to modern transmitters came in several steps. First came continuous wave oscillators, which offered stable, narrowband carriers for the on-off modulation used in radiotelegraphy. Next, the stability of these oscillators made amplitude modulation (AM) practical, so voice could now be transmitted wirelessly. However, AM bandwidths are considerably wider than those required for keyed continuous wave (CW) transmissions.

A clever approach to sending the same information in half the bandwidth, and with less overall power, was developed by Carson [7]. Figure 1.2 shows the situation in the frequency domain. In Figure 1.2(a), we see the spectrum of an unmodulated radio carrier: all of the power in the transmission is concentrated at the carrier frequency. Figure 1.2(b) shows the result of amplitude modulating that carrier: some of the energy is now found symmetrically distributed above and below

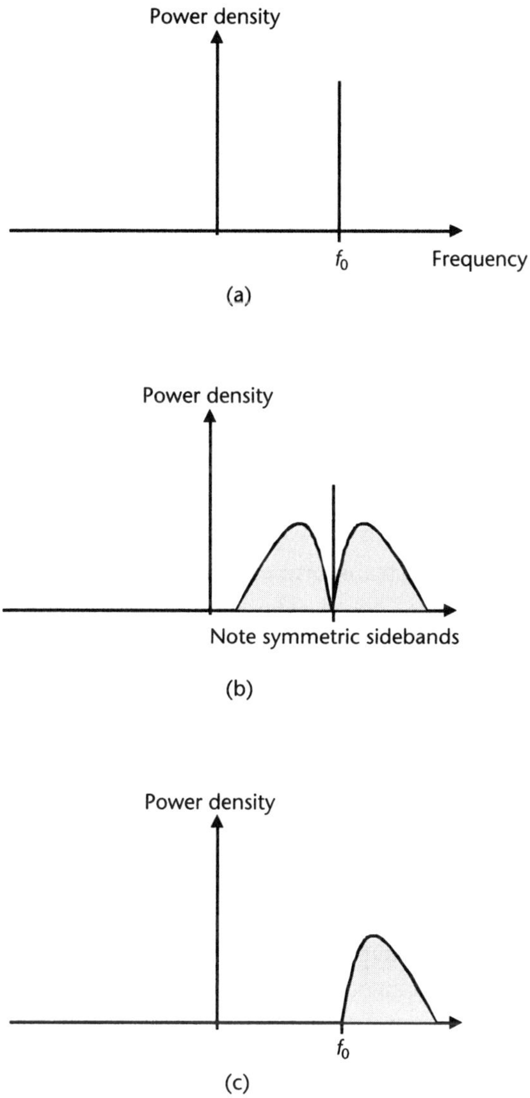

Figure 1.2 Single sideband.

the carrier frequency[2]. These adjacent portions of the spectrum that carry energy resulting from modulation are termed sidebands, and it is here that the desired information is to be found. The information in the two sidebands is identical (although the lower sideband is the mirror image of the upper). We can conserve both bandwidth and transmitter power if we send only one of the sidebands, filtering out the carrier and the other sideband. This single-sideband (SSB) mode of operation, shown in Figure 1.2(c), is now the de facto standard in the HF band.

2. For example, the mathematical representation of 100% amplitude modulation of a carrier of frequency, f_c, by a single tone of frequency, fm, shows how these symmetric sidebands arise:

$$[\cos(2\pi f_m)+1]\cos(2\pi f_c)=\cos(2\pi f_c)+\frac{1}{2}\{\cos[2\pi(f_c+f_m)]+\cos[2\pi(f_c-f_m)]\}$$

Modern HF radiotelephone service provides reasonable voice quality in a 3-kHz channel by filtering the voice signal to a 3-kHz bandwidth and employing SSB transmission. As a result, spectrum in the HF band is normally allocated in 3-kHz channels, and HF modems have therefore evolved to work within this narrow bandwidth (as described in Chapter 3).

1.2 HF Antennas

Unless an antenna is electrically small [8], its physical size will be comparable to the wavelength of the radio waves it handles. With wavelengths ranging from 10 to 150m, HF antennas can be quite large, and are often the most visible elements of an HF radio transmission system (see generic block diagram in Figure 1.3).

1.2.1 Transmitting Antennas

Transmitting antennas transform electrical energy from the transmitter into propagating electromagnetic waves. Despite the better long-range propagation offered by the skywave channel (see Chapter 2), compared to the surface-wave channel used by Marconi et al. [1], long-haul HF transmitters still operate at power levels of thousands of watts, and the transmitting antenna must be able to handle these high voltages and currents.

Such high power levels also increase the importance of matching the radio frequency (RF) impedance of the antenna to that of the transmitter. An impedance mismatch results in power being reflected from the antenna back to the transmitter; severe mismatches can result in arcing, danger to personnel, and the destruction of transmitter components.

As we will see in Chapter 2, not all frequencies within the HF band will propagate to the desired receiver location, so HF antennas must be able to operate over a wide range of frequencies. This complicates the requirement to match the RF impedance of the antenna to that of the transmitter at the current operating frequency. In particular, small HF antennas generally do not offer a good impedance across the HF band, so we must insert reactive elements (coils and capacitors) between the transmitter and the antenna to improve the impedance match. Such *antenna couplers* (also known as *antenna tuners*) were once manually adjusted, requiring many

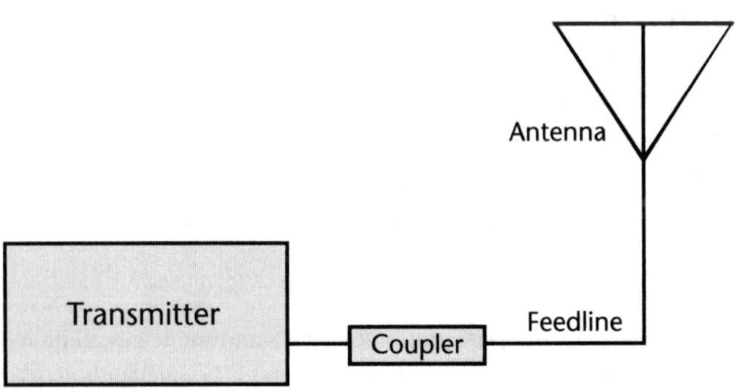

Figure 1.3 HF transmission system.

seconds to tune. Modern couplers are often under microprocessor control and can remember the settings needed for each operating frequency, thereby completing their tuning in a fraction of a second.

1.2.2 Receiving Antennas

HF receiving antennas are often quite different from transmitting antennas. Superficially, they perform the inverse function: intercepting ambient electromagnetic waves and converting them to electrical energy. However, the power levels are far lower than those at the transmitting antenna, and the impact of an impedance mismatch is far lower. As we will see in Chapter 2, the dominant sources of noise for an HF receiver lie outside of the radio. Therefore, an impedance ismatch between the receiving antenna and the receiver usually results in equal losses of incoming signal and noise, so the net signal-to-noise ratio is unchanged.

1.2.3 Antenna Polarization

Polarization of an electromagnetic wave refers to the orientation of the electric field component with respect to the Earth's surface. HF antennas may be designed to emit waves that are vertically, horizontally, or even circularly polarized.

- Vertically polarized antennas, such as whips, are typically omnidirectional in azimuth with a null overhead.
- Horizontally polarized antennas include simple antennas, such as horizontal dipoles, as well as log periodic and other large broadband antennas. They are often directional in azimuth and have strong emissions overhead.

Good references for further reading in HF antennas include the *ARRL Antenna Book* [9] for an introduction to antenna design, and Johnson et al. [8] for detailed discussions of a wide range of HF antennas.

1.3 HF Radio in the Computer Age

The advent of ubiquitous computing has brought profound changes to HF radio, arguably as significant as the shift a century ago from longwave spark transmitters to SSB HF. The impact of computing on HF radio began with a demand for data modems for HF channels (Chapters 3 and 6). In the 1980s, microprocessors were found useful in automating the operations of HF radio communications (Chapter 4), and we are now in the third generation of such automatic link establishment (Chapter 5). Standardization has been the key to rapid adoption of each new generation of HF technology [8].

1.4 Summary

HF radio offers an alternative to satellites for beyond-line-of-sight wireless communications, thus avoiding the costs, vulnerabilities, and sovereignty concerns of satellite communications. Chapter 2 begins our technical discussions with a review of how HF radio signals propagate via the ionosphere. The following chapters then explore recent developments in HF technologies for harnessing skywave channels for voice and (especially) data communications.

References

[1] Aitken, H., *Syntony and Spark–The Origins of Radio*, New York: Wiley and Sons, 1976, p. 143.

[2] ibid., p. 268.

[3] ibid., p. 272.

[4] "Recommendations Of The National Radio Committee," *Radio Service Bulletin*, U.S. Department of Commerce, April 2, 1923.

[5] Hong, S., *Wireless, from Marconi's Black-Box to the Audion*, Cambridge MA: MIT Press, 2001, p. 101.

[6] ibid., p. 110.

[7] Carson, J., /AT&T, "Method and Means for Signaling with High Frequency Waves," U.S. Patent 1,449,382 filed December 1, 1915, granted March 27, 1923.

[8] Johnson, E., et al., *Advanced High-Frequency Radio Communications*, Norwood, MA: Artech House, 1997.

[9] *ARRL Antenna Book*, 22nd Edition, Newington, CT: American Radio Relay League, 2011.

CHAPTER 2

The HF Channel

The unique capabilities of HF radio come with a correspondingly unique set of challenges. The key to achieving the potential of low-cost, over-the-horizon communications lies in understanding the physics of ionospheric propagation and the resulting statistical properties of the HF skywave channel. The purpose of this chapter is to provide an overview of these topics. For the reader who is interested in more in-depth coverage of this area, a number of worthwhile books are available, including those by Goodman [1], Maslin [2], and Johnson et al. [3].

2.1 Surface-Wave Propagation

Before we immerse ourselves in the arcane aspects of radio wave propagation via ionospheric refraction, it is important to remember that HF radio waves also propagate by line-of-sight (familiar to users of higher frequencies) as well as surface-wave paths. Such propagation is *nondispersive*[1], and is therefore free of the type of fading common to skywave channels (this is discussed in Section 2.2.4). A signal received via a surface-wave path will be an "attenuated, delayed, but otherwise undistorted version of the transmitted signal" [2]. Maslin provides a useful rule of thumb for attenuation versus distance for surface-wave operation, using three zones:

- In the *direct radiation zone*, the power density falls proportional to the distance squared.
- At distances beyond the direct radiation zone, we can use *Sommerfeld flat earth theory* [4], in which power density falls proportional to the distance to the fourth power.
- Beyond the *Sommerfeld zone*, we enter the *diffraction zone*, where power density falls exponentially with distance.

In practice, surface-wave propagation is useful over ranges up to a few hundred nautical miles over sea water, but is restricted to shorter distances over land.

1. The speed of wave propagation is independent of frequency.

2.2 Skywave Propagation

Surface-wave operation offers over-the-horizon communication, but its range is certainly not global. To extend our reach to thousands of kilometers, we must take another approach: launching our signals skyward, where they will interact with the ionosphere and return to Earth far beyond the horizon. We therefore turn now to a study of this curious outer region of Earth's atmosphere.

2.2.1 The Ionosphere

In addition to producing visible light, the sun radiates throughout the electromagnetic spectrum, including torrents of ionizing radiation (far ultraviolet, X-rays, and gamma rays). The Earth's magnetic field deflects much of this *solar wind* (see Figure 2.1), especially in the equatorial and temperate regions. At the Earth's surface, the intensity of ionizing radiation is very low (which is fortunate for the safety of those living here). However, the effectiveness of Earth's "deflector shield" decreases with distance above the surface. At the outer reaches of the atmosphere, the intensity of solar radiation is much higher, and the few gas molecules found there are frequently ionized. That is, one or more of their electrons may be liberated in a collision with an energetic photon from the solar wind. These free electrons eventually recombine with an ion, but can remain in the ionized plasma for some time.

The region of the atmosphere having significant free electron density is termed the *ionosphere*, which is generally considered to lie between about 70 and 600 km above Earth's surface. The electron density varies with height:

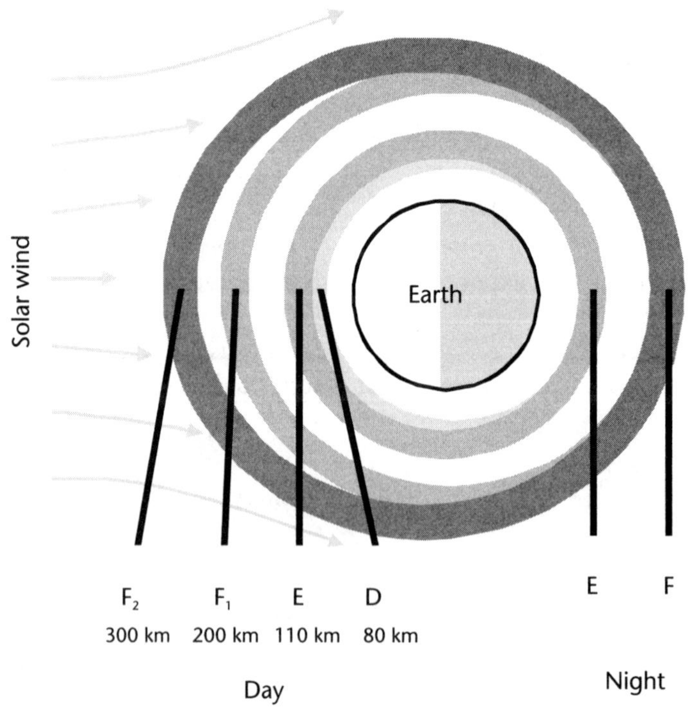

Figure 2.1 Solar wind and ionospheric layers.

- At the lower altitudes, the electron density is low due to low intensity of ionizing radiation (only the most energetic photons penetrate this far) as well as high neutral gas density. The latter results in frequent collisions with the free electrons and rapid recombination.
- At the highest altitudes, the electron density is also low. This is due to the low density of gas molecules to be ionized.
- The highest electron density is found in the vicinity of 300 km altitude [1]. Here the gas density is very low, with little ionization taking place. However, free electrons that diffuse upward from lower altitudes have relatively long lifetimes due to few opportunities for recombination.

The dominant gas species (and ionization mechanism) vary with altitude [1], with the heaviest gas molecules found at the lowest altitudes. This stratification of ionized species contributes to the layered structure of the ionosphere shown in Figure 2.1:

- The *D layer*, lying between about 70 and 90 km in altitude, has low electron density and high neutral gas density. It is formed mostly from the ionization of nitric oxide by gamma and X-rays. The D layer is strongest during the day, when the solar radiation is present. At night, cosmic rays produce ions here, but at a very low rate.
- The *E layer*, from about 90 to 120 km, has higher electron density and lower neutral gas pressure than the D layer. It is largely formed by solar ionization of molecular oxygen.
- The *F layer* is the outermost of the three, and is found from about 200 km up to 500 or 600 km. The F layer has the highest electron density and the lowest neutral gas density. At this altitude we find the lightest gas molecules: hydrogen, helium, and monatomic oxygen. During the day, intense solar radiation creates a two-layer structure in the F region. The F_1 layer is the lower of the two, and is present only during the day. The daytime F_2 layer persists through the night, when it is labeled just the "F layer."

Even within a layer, the electron density is not uniform, but varies with altitude. Each layer in Figure 2.1 is labeled with the approximate altitude of the peak ionization of that layer.

The rate of ion production in any volume of the ionosphere varies with the overall intensity of solar ionizing radiation impinging on that volume of gas according to several physical processes:

- The intensity of solar radiation is highest at local noon each day. Solar radiation vanishes at night, and electrons liberated during the day gradually recombine until sunrise. This daily variation is termed the *diurnal cycle*.
- The apparent elevation of the sun varies with the season: the hemisphere experiencing summer receives more direct illumination for a longer period each day than does the other hemisphere.

- The overall intensity of radiation from the sun varies over a roughly 11-year period, correlated with the sunspot cycle.
- Various aperiodic events, such as solar flares and coronal mass ejections, produce large bursts of ionization. These tend to occur more often near the peak of the sunspot cycle than at its minimum.

The periodic mechanisms produce predictable variations in the ionosphere that are well captured in propagation prediction programs (see Section 2.4), but the aperiodic events can be troublesome. Another element of variability, not attributable to solar activity, is the occasional appearance of strongly reflective *sporadic E* patches, which are considered to be monatomic metallic ions from meteoric debris collected into patches by wind shear [1].

2.2.2 Ionospheric Propagation

The free electron density of an ionized region affects the phase velocity [1] of electromagnetic waves passing through it. Because the electron density varies with height in the ionosphere, electromagnetic waves will be refracted (bent) as they pass through regions with significant ionization gradients. This effect varies with the frequency of the wave: as the frequency increases, the amount of refraction decreases for a given ionization profile.

The frequency dependence of ionospheric refraction leads to the notion of a *critical frequency*. A region having free electron density N (in electrons per m^3) will effectively *reflect* radio waves below the critical frequency f_0 (in Hz):

$$f_0 = 9\sqrt{N}$$

That is, a radio wave at a frequency f_0 traveling vertically until it reaches an ionized region with free electron density N will be refracted through 90° as it ascends to that region, and will then be refracted symmetrically through an additional 90°, returning the wave to its source. Waves at frequencies greater than f_0 are refracted less than 180° and are therefore not returned to the source (but might be received at locations distant from the source). Some of the energy in an incident radio wave is absorbed in the ionized region, rather than refracted; however, as the frequency increases the attenuation due to absorption decreases.

We employ ionospheric refraction for long-range communication by "bouncing" radio waves from the ionosphere at grazing angles. This permits the use of frequencies higher than the critical frequency because we do not require 180° of refraction. For a geometry with zenith angle ϕ (see Figure 2.2), the oblique critical frequency $f_{max}(\phi)$ can be computed using a secant law:

$$f_{\max}(\phi) = f_0 \sec(\phi)$$

As noted above, ionization varies widely with time, season, and solar activity, but we can gain some sense of the range of usable skywave frequencies from typical order-of-magnitude values for the peak electron density of the ionospheric layers:

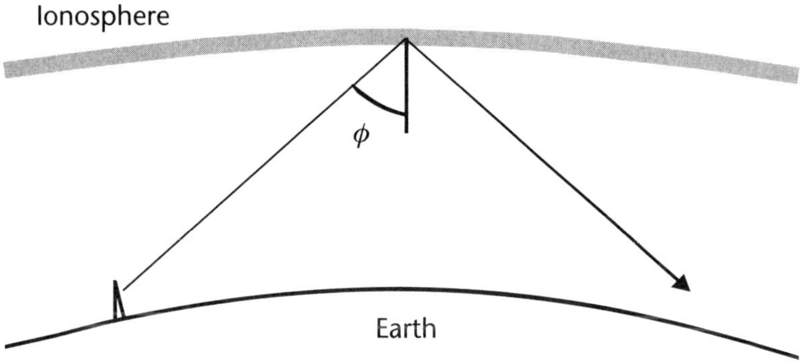

Figure 2.2 Oblique refraction.

- Peak free electron density in the daytime D layer may reach 10^9 electrons per m^3 with a critical frequency of about 300 kHz. This critical frequency is far below the HF band, so all HF waves will pass through the D layer with little refraction. However, frequencies at the lower end of the HF band will experience significant attenuation when passing through the daytime D layer. Thus, the D layer sets a lower limit on usable frequencies during the day for long-range communications.
- In the E layer, daytime free electron density peaks at roughly 10^{11} per m^3 with a critical frequency of about 3 MHz, near the bottom of the HF band. At oblique angles, the E layer can be used for medium-range communications.
- The F_2 layer may have a peak free electron density of up to 10^{12} per m^3 resulting in a critical frequency of about 9 MHz, with oblique refraction possible over much of the HF band. At night, the peak ionization decays slowly by one to two orders of magnitude. Since the F layer has the highest free electron density in the ionosphere, frequencies that breach the F layer continue into space. Thus, the critical frequency of the F layer sets the upper bound on usable frequencies for a path. Due to its height, reflections from the F layer have ranges of thousands of kilometers, so the F layer is generally preferred for long-range skywave communications.

To communicate over distances greater than the few thousand kilometers possible with a single reflection from the F layer, we must employ a multihop path. In this case, signals returning to Earth from the ionosphere are scattered from the Earth's surface. Some portion of that energy travels upward to reflect again from the ionosphere. Note that each reflection or scattering results in significant loss of signal strength. Skywave paths often have end-to-end losses exceeding 100 dB.

2.2.3 Near Vertical Incidence Skywave

As we have seen, the ability to bounce radio waves from the ionosphere supports communications far beyond the horizon, with ranges of thousands of kilometers. Sometimes skywave propagation also solves short-range communications problems. For example, when reaching a nearby site is obstructed by terrain or buildings, we can transmit HF radio signals nearly straight up, where they will be returned by the ionosphere to nearby receivers, nicely hopping over the obstructions. This mode of operation is termed near vertical incidence skywave (NVIS).

Because of the nearly vertical angle of incidence on the ionosphere, NVIS operation requires the use of lower frequencies than long-haul skywave. Also, because of the D layer absorption of these low frequencies during the day, NVIS may require more power than might be expected.

2.2.4 Skywave Fading

So far, we have described an idealized ionosphere. The real ionosphere is a dispersive medium. Signals propagated through the ionosphere are spread in both time and frequency[2]. This spreading is due to variations in phase velocity throughout the plasma, and to motion of different parts of this tenuous region of the atmosphere. This distorts analog signals that are carried by HF transmissions, and introduces intersymbol interference[3] in data waveforms (see Chapter 3).

In addition to the very slow variations in skywave signal strength due to the diurnal, seasonal, and sunspot cycles, there are numerous features of ionospheric propagation that result in faster fluctuations and fading, described in Sections 2.2.4.1, 2.2.4.2, and 2.2.4.3. These sources of fading are discussed in more detail in [1].

2.2.4.1 Multipath Propagation

A single signal usually propagates to the receiver via many paths through the ionosphere. Because the path lengths differ, the arriving signals have different phases and interfere with each other. Fading results as the relative phase of the interfering signals varies, as illustrated in Figure 2.3. This figure shows a transmitted signal, a second version shifted slightly in frequency, and the sum of the two, which arrives as a fading signal at the receiver. In many cases, the received signal will be a superposition of more than two copies of the signal, with time-varying phase relationships.

Such multipath fading in long-haul skywave channels is often modeled as Rayleigh-distributed[4] because there is no dominant component in the received signal. (Rician fading results when energy arrives at a receiver via both ionospheric and surface-wave paths.)

The duration of fades depends on the scale of the path differences.

- *Micromultipath* arising from motion of small inhomogenieties in the ionosphere gives rise to *flutter fading* with periods of less than 1 second.

2. That is, energy emitted at a certain instant and at a certain frequency may arrive at a receiver at multiple times and frequencies, with either discrete or continuous distributions.
3. For example, if we have 2 ms of spreading in the time domain and the modem symbol duration is 417 μs, echoes of one symbol may corrupt the following four symbols.
4. The arriving signal may be analyzed in its in-phase and quadrature components, with the energy arriving via each path contributing to both components. If the received signal is the result of combining many small contributions, the components will have independent Gaussian distributions (by the central limit theorem), which means that the amplitude of the received signal will be Rayleigh-distributed.

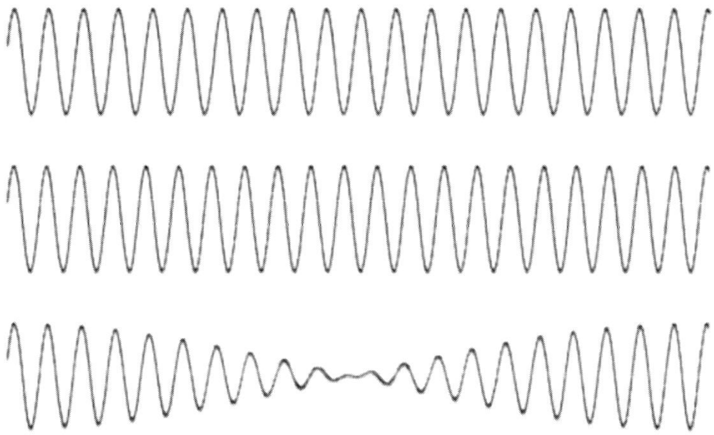

Figure 2.3 Fading example.

- Multipath arising from multihop or multilayer paths may produce slower fading on the order of a few seconds [1].

In general, fading resulting from the relative motion of refracting regions tends to be more rapid at the upper end of the HF spectrum, since a given displacement of the refracting region corresponds to a greater phase shift for shorter wavelengths.

2.2.4.2 The Faraday Effect

Another source of fading is the Faraday effect, in which the polarization of a refracted signal slowly rotates, resulting in fades from seconds to minutes in duration [1].

2.2.4.3 Ray Interference

Lens-type irregularities form in the ionosphere that result in slowly varying ray interference. These fades have periods of tens of minutes [1].

In short, the transfer function of skywave paths may be considered as nonstationary on nearly any time scale. This is certainly a challenge to communication engineers!

2.3 Noise in the High-Frequency Band

In the high-frequency (HF) band (and below), noise levels from external sources are so high that they normally overwhelm internal thermal noise in the receiver[5], with antenna noise factors of 30 to 70 dB [3]. This noise arises from three sources: human activity (manmade noise), terrestrial lightning (atmospheric noise), and ce-

5. When the receiving antenna is extremely inefficient (e.g., buried), the external noise may be attenuated below the receiver noise floor. In such cases, an external low-noise amplifier may be employed to preserve the ambient signal-to-noise ratio.

lestial radio sources (galactic noise). The noise produced by all of these sources can propagate to the receiver either directly or via surface-wave or skywave paths.

In general, noise from these external sources is most intense at the lower frequencies, with steadily falling noise power as the operating frequency increases. This is due both to the characteristics of the noise source and to the filtering effect of skywave propagation.

The noise in the HF band is mainly impulsive, but we often model it as additive white Gaussian noise (AWGN) for mathematical simplicity. This modeling convenience can result in higher modem error rates in AWGN channel simulations than when using measured HF noise [3].

2.4 Models of the HF Communication Channel

The ionospheric channel is well-known for exhibiting temporal effects over a wide range of time scales [1], including multipath spreads on the order of milliseconds that produce intersymbol interference in modems, fading on the order of seconds to minutes, hourly diurnal variations, and so on up through the 11-year sunspot cycle. Despite these channel impairments, the value of beyond-line-of-sight wireless communications is such that technologies have been developed to deal with each of these challenges. These techniques are often evaluated via simulation in the earliest stages of technology development, so the HF radio research community has found it useful to agree on standard approaches to simulating the ionospheric channel.

In this section, we will discuss modeling the ionospheric channel as a superposition of effects due to three categories of considerations:

- Space weather, signal path geometry relative to the sun, and other slowly varying factors;
- Fading effects due to ionospheric motion, Faraday rotation, and similar phenomena (intermediate time scales);
- Multipath interference, which produces Rayleigh or Rician fading (shortest time scales).

The first category of effects has been captured in the models used in well-known ionospheric propagation prediction programs, such as VOACAP [5] and ICEPAC [6]. The third category is usually represented using the Watterson model [7]. The intermediate category is not as well-known, however, so it is described in some detail below.

2.4.1 Propagation Prediction

The periodic diurnal and seasonal cycles are well-understood, and their effects are predictable using little more than spherical trigonometry. However, local variations in ionospheric propagation depart from the simple models. These effects have been measured over the years, resulting in maps of correction coefficients that can be applied to the geometric models. Space weather conditions, including the sunspot number, are usually provided explicitly as an input to prediction programs.

HF communication engineers often use a prediction program to evaluate the feasibility of ionospheric circuits and to identify frequencies likely to propagate under the conditions of interest. Using VOACAP [5] as an example, the user provides the following data as input to the program:

- Transmitter location (latitude and longitude), power, and antenna model;
- Receiver location, antenna model, and local noise;
- Month(s) and hours of interest;
- Smoothed sunspot number (SSN);
- Frequencies of interest.

The program then computes for each frequency of interest:

- The viable paths (modes) that a signal can take through the ionosphere from the transmitter to the receiver;
- The losses encountered along those paths;
- The resulting signal strength arriving at the receiver (taking into account the antenna gains at the computed azimuths, takeoff, and arrival angles);
- The signal-to-noise ratio (SNR) relative to the estimated noise at the receiver at that frequency.

These results can be presented in tabular form or plotted graphically. An example of the latter is shown in Figure 2.4, where we see contours of signal-to-noise density[6] as a function of hour (Universal time) and frequency. The propagation prediction is for the month of March for a circuit in the central United States, with a sunspot number of 47. The control point of this circuit (i.e., the midpoint of the single-hop path, where the radio waves interact with the ionosphere) lies six time zones to the west of the Prime Meridian, so local noon appears at hour 18 in the graph.

In this graph, we can clearly see the effects of the diurnal cycle that were discussed in a general way above:

- At noon (hour 18), we have the highest ionization throughout the ionosphere, which supports use of frequencies up to about 19 MHz. Also, the D layer is strongest at noon, and its attenuation renders frequencies below about 11 MHz unusable.
- At sunset (hour 24 on Earth's surface, later at the altitude of the ionosphere), the production of free electrons by the sun ceases, and the critical frequency begins to drop rapidly as recombination reduces the electron density in the E and F layers. Happily, the D layer also dissipates, so lower frequencies become usable.

6. SNR in a 1-Hz bandwidth. To convert to SNR in a 3-kHz bandwidth, we need to subtract $10 \log_{10}(3000)$ or approximately 35 dB.

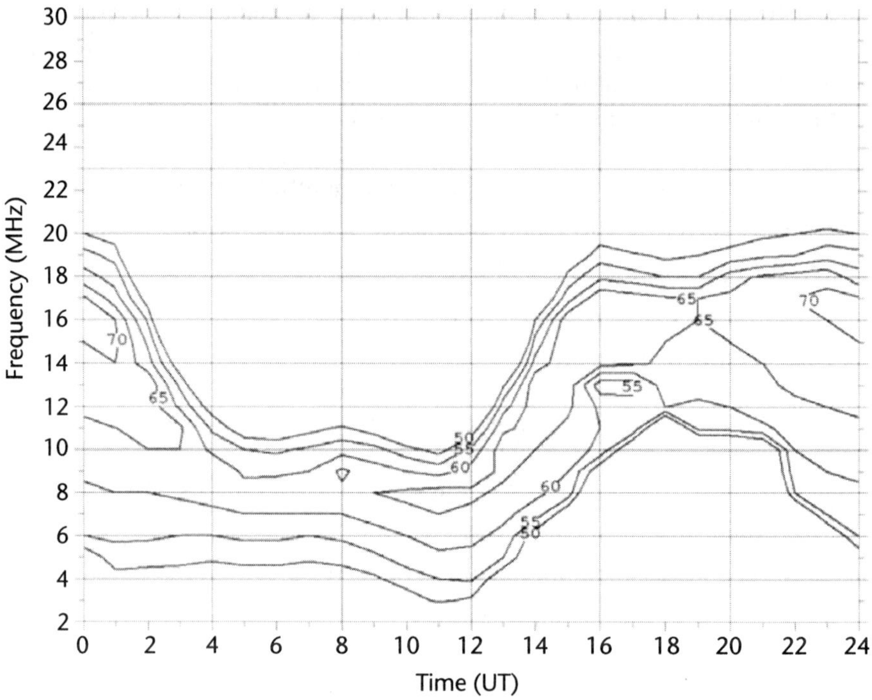

Figure 2.4 Example VOACAP SNR graph.

- Through the night (roughly hours 4 through 10), residual free electrons in the F layer support communications, but on lower frequencies than those usable during the day.
- At sunrise, we see ion production begin again, and the band of usable frequencies climbs back up to the daytime range during the early morning hours.

In this case, VOACAP predicts that the 3-kHz SNR throughout the day would be at least 25 dB on the best frequencies, so this circuit should be usable for voice or high-speed data (Chapter 3).

In addition to monthly median SNR values, VOACAP and similar programs also estimate the statistical spread of SNR values to be expected during the month. These are provided as the first and ninth deciles (which are usually asymmetric with respect to the median).

2.4.2 The Watterson Model

Propagation prediction programs are able to provide broad indications of the usability of an ionospheric path at the granularity of one hour, and averaged over one month. While these long-term, statistical predictions are quite valuable for engineering an HF circuit, the designers of HF data communications technologies require models of the fine-grain behavior of ionospheric paths on a scale of microseconds to milliseconds.

It is immediately apparent to anyone communicating over an HF skywave path that the signal fades frequently and sometimes deeply. The model generally used for the randomly varying amplitude and phase of a skywave signal is Rayleigh fading[7]. In the late 1960s, Watterson and his colleagues determined that a relatively simple mathematical model exhibits most of the features of measured skywave fading, and published what has since become known as the Watterson model [7].

While recognizing that the skywave channel is nonstationary in both time and frequency, Watterson et al. proposed a stationary model that adequately approximates the real channel over a bounded bandwidth. Their model is now widely accepted, and even mandated for use in testing HF modems [8]. Although simulators that implement this model do not exhibit exactly the same behavior as the real channel, we have found that modems that perform well on a Watterson model channel simulator will also perform well in the real channel [9].

When used for testing HF modems, the Watterson model is simplified to treating a signal as traveling via exactly two paths, with equal average path losses [8]. The amplitudes on each path vary independently according to fading-gain processes with a common Gaussian power spectral density. The differential time delay between the two paths is fixed. At the output of the simulator, the two independently fading signals are combined with additive white Gaussian noise.

Three parameters thus determine the behavior of a Watterson channel simulator:

- Time delay between the two paths (delay spread);
- Bandwidth of the fading-gain processes (Doppler spread);
- Signal-to-noise ratio (averaging the summed power of the two fading signals, relative to the noise, over a very long period of time).

Representative values for the fading parameters (shown in Table 2.1) may be found in the ITU-R recommendation for using Watterson channel simulators to test HF modems [8]. These testing parameters assume two equal power paths with a fixed delay between the paths defined by the delay spread. Each path exhibits Rayleigh fading with a fading bandwidth defined by the Doppler spread.

A study by a subcommittee of the HF Industry Association determined that imprecise implementation of the Watterson model in HF channel simulators was resulting in inconsistent results when different simulators were used in testing [10]. The recommendations of this subcommittee led to standardized specifications for simulators used to test compliance with military HF modem standards [11].

Table 2.1 Mid-Latitude Channel Characteristics from ITU-R F.1487

Channel Condition	*Delay Spread (ms)*	*Doppler Spread (Hz)*
Quiet	0.5	0.1
Moderate	1	0.5
Disturbed	2	1
Disturbed NVIS	7	1

7. When the received signal includes both surface-wave and skywave components, the fading process is better modeled as Rician because there is a single strong signal, plus many small contributions, from the skywave.

2.4.3 Midterm Variation

Watterson model channel simulators are now commonly used in the design, analysis, and performance evaluation of HF modems and of communications protocols at the link, network, transport, and even application layers. However, the protocols (which exchange packets in both directions on a link over extended periods) are sensitive to effects at the seconds-to-minutes scale that are present in the ionosphere but not represented in the Watterson model. As noted in Section 2.2, these intermediate-scale ("midterm") variations in signal strength arise from ionospheric motion, Faraday rotation, and focusing by ionospheric irregularities. Statistically, the intermediate-term variations seem to be lognormally distributed [1]. That is, the signal strength in decibels has a normal distribution.

Furman and McRae [12] measured these variations on a link from Rochester, New York, to Melbourne, Florida. They confirmed the lognormal distribution of variations in SNR, and measured a standard deviation of about 4 dB. The autocorrelation of these variations was found to be approximately exponential with a time constant of around 10 seconds.

Johnson and colleagues incorporated these findings into the *Walnut Street model* [13], a standard approach for modeling ionospheric channels in long-running simulations of HF networks (i.e., hours and days of simulated time). This model combines the effects of the three regimes as follows:

- A validated propagation model, such as VOACAP, is used to predict the statistics of SNR on a link for each hour. A random quantile is chosen for each simulation, and the SNR value at that quantile is recorded for each hour. (That is, if the SNR for hour 0 falls at the 70th percentile of the hourly distribution for that hour, then the SNR for 0100 will be the 70th percentile of the hourly distribution for hour 1, and so on. Subsequent simulations would use different quantities.) The long-term average SNR process at any instant during the simulation is computed by interpolating between the hourly values.
- The midterm variation in SNR is generated as a lognormal random process whose standard deviation and time constant are specified for the simulation (e.g., using the findings of Furman and McRae [12]: a 4-dB standard deviation with a 10 s time constant). The midterm variation is added to the interpolated long-term average SNR value (as shown in Figure 2.5).
- The instantaneous SNR values resulting from this procedure are used by modem models based on Watterson channel simulations, thereby capturing the short-term Rayleigh fading.

This approach has been used successfully in large-scale HF network simulations for many years, and has also been used to explore potential interactions of ionospheric fading with HF data link protocols [14].

More recently, frequency-domain analysis of intermediate-term fading measurements by Batts, et al., [15] has found that these fades are better modeled as *two* lognormal processes:

- A slow, large-amplitude process termed *long-term variation* (LTV).

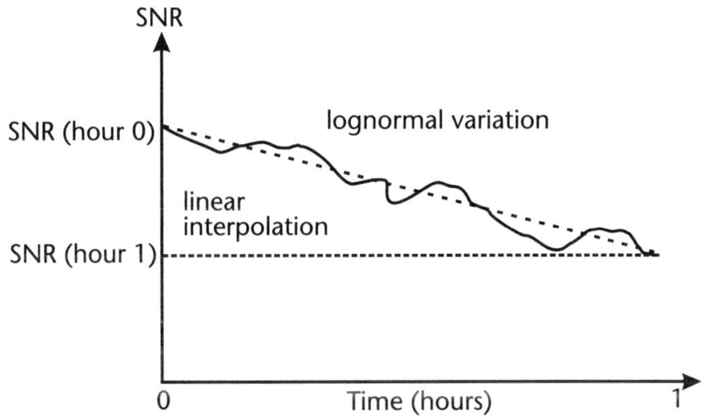

Figure 2.5 Walnut Street example.

- A faster but lower amplitude process termed *intermediate-term variation* (ITV).

In this study, a series of SNR measurements at 2.7 s intervals was high-pass filtered with a cutoff frequency of 0.001 Hz to remove diurnal variations. When fast Fourier transforms (FFTs) were computed from the filtered data, a large peak was found between 0.001 Hz and 0.01 Hz. This corresponds to the LTV component. When this was removed from the data, the residual variation had a shorter time constant, which was termed ITV.

The orignal Walnut Street modeling approach is easily generalized to incorporate two lognormal components, as shown in Figure 2.6. Each of the lognormal components is characterized by two parameters:

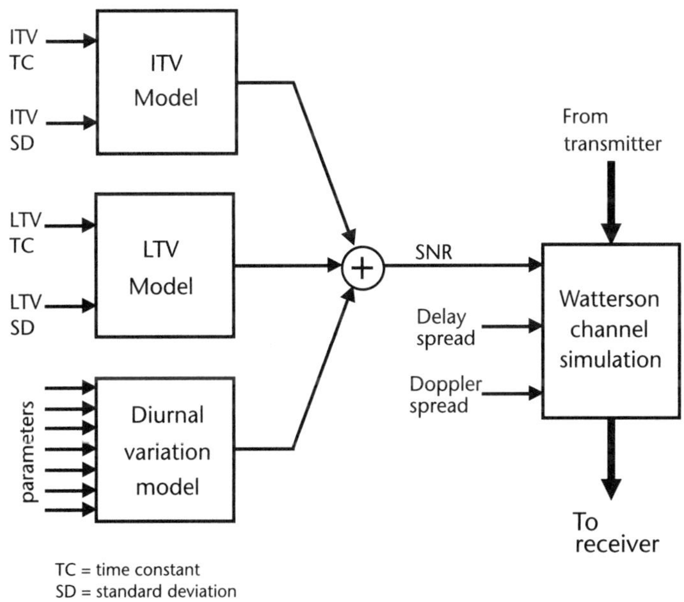

Figure 2.6 Generalized Walnut Street model.

- The standard deviation (SD) of the lognormal distribution determines the amplitude of the variation in decibels.
- The time constant (TC) of the exponential autocorrelation determines the fading rate.

This model has been fitted to measurements of two skywave paths [15]:

- A long-haul path from Rochester, New York, to Palm Bay, Florida (1697 km).
- A NVIS path from Rochester to Wolcott, New York (61 km).

The parameters in Table 2.2 provided very good agreement between the measured SNR spectra and those produced by the model (i.e., differences in cumulative spectrum profiles were generally less than 1%) [15].

Table 2.2 ITV and LTV Parameters of Measured Paths

Path	*ITV SD (dB)*	*ITV TC (s)*	*LTV SD (dB)*	*LTV TC (s)*
Long-haul	3.9	5.2	3.5	180
NVIS	2.3	1.9	1.8	97

2.5 Summary

An HF skywave channel conveys signals beyond line-of-sight via refraction in the ionosphere to one or more distant receivers. (Sometimes a signal is refracted by the ionoshpere and scattered from Earth's surface multiple times on the path from transmitter to receiver.) The refractive and absorptive characteristics of the ionospheric layers depend strongly on radio frequency, latitude, time of day, season, the solar weather, and so on.

Signals reach the receiver via refraction from one or more ionospheric layers, each of which may be in motion. The received signal is often a composition of multiple signals having independent, time-varying path losses and phase shifts. Thus we may expect multipath interference, deep fades, and impulsive (non-Gaussian) noise, all superimposed on a slowly varying SNR trend. In the next chapter, we explore the techniques used in HF modems to pass data via this challenging channel.

References

[1] Goodman, J., *HF Communications: Science and Technology*, New York: Van Nostrand Reinhold, 1992.

[2] Maslin, N., *HF Communications: A Systems Approach*, London: Plenum Press, 1987.

[3] Johnson, E., et al., *Advanced High-Frequency Radio Communications*, Norwood, MA: Artech House, 1997.

[4] Sommerfeld, A., "The Propagation of Waves in Wireless Telegraphy," *Ann. Phys., Series 4*, Vol. 28, 1909, pp. 665.

[5] Perkiömäki, J., "VOACAP Quick Guide," http://www.voacap.com/index.html (last accessed March 2012).

[6] Stewart, F. G., "Technical Description of ICEPAC Propagation Prediction Program," http://www.voacap.com/itshfbc-help/icepac-tech-intro.html (last accessed March 2012).

[7] Watterson, C., J. Juroshek, and W. Bensema, "Experimental Confirmation of an HF Channel Model," *IEEE Transactions on Communication Technology*, Vol. COM-18, No. 6, 1970. (Also published as "Technical Report ERL 112-ITS 80, Experimental Verification of an Ionospheric Channel Mode," U. S. Department of Commerce Environmental Science Services Administration, Boulder, CO, 1969.)

[8] ITU-R Recommendation F.1487, "Testing of HF Modems with Bandwidths of up to About 12 kHz Using Ionospheric Channel Simulators," *International Telecommunication Union, Radiocommunication Sector*, Geneva, 2000.

[9] McRae, D., and F. Perkins, "Digital HF modem performance measurements using HF link simulators," *Fourth International Conference on HF Radio Systems and Techniques*, London, UK: IEEE 1988.

[10] Furman, W., and J. Nieto, "Understanding HF Channel Simulator Requirements in Order to Reduce HF Modem Performance measurement Variability," Proceedings of the 2001 Nordic Shortwave Conference (HF01), Fårö, Sweden, 2001.

[11] MIL-STD-188-110C, Appendix H, "Characteristics of HF Channel Simulators," DoD, 23 September 2011.

[12] Furman, W., and D. McRae, "Evaluation and Optimization of Data Link Protocols for HF Data Communications Systems," *Proceedings of 1993 Military Communications Conference*, Boston, MA, October 1993.

[13] Johnson, E., "The Walnut Street Model of Ionospheric HF Radio Propagation," *NMSU Technical Report*, May 1997.

[14] Johnson, E., "Interactions Among Ionospheric Propagation, HF Modems, and Data Protocols," *Proceedings of the 2002 Ionospheric Effects Symposium, IES '02*, Alexandria, VA, May 2002.

[15] Batts, W., Jr., W. Furman, and E. Koski, "Empirically characterizing channel quality variation on HF ionospheric channels," *Proceedings of the 2007 Nordic Shortwave Conference* (HF 07), Fårö, Sweden 2007.

CHAPTER 3

Data Transmission in 3-kHz Channels

Data communication over ionospheric channels has been evolving since the very beginnings of HF radio. The first HF signals—based on spark gap transmitters—quickly developed into more sophisticated communications networks, though they still used human operators and Morse code alphabets. Improvements, such as more complex modulation techniques—including binary FSK, *M*-ary FSK, parallel-tone, and orthogonal frequency division multiplexing (OFDM) modulations—rapidly ensued. Although OFDM is currently used in commercial HF broadcast applications, such as Digital Radio Mondiale (DRM), today's high-performance, two-way data communications (such as military communications) generally employ serial-tone (i.e., single-carrier) waveforms with robust forward error correction schemes, operating in 3-kHz channel allocations. As a class (when demodulated with sophisticated equalization techniques) these waveforms offer significantly better all-around performance over HF radio channels than any previous class of modulation.

3.1 Introduction

As noted in Chapter 2, the HF skywave channel may be considered nonstationary on nearly any time scale. Any attempt to send data over such paths must address the time-varying, dispersive nature of this channel at appropriate points in the transmitting and receiving systems:

- Long-term variations are best addressed by selecting a suitable operating frequency, a task that modern networks relegate to automatic link establishment (ALE), discussed in Chapter 4.
- Intermediate-term instabilities (on the order of seconds to minutes) can be overcome by a data-link protocol that repeats lost frames upon request.
- The shortest-term effects (noise, multipath, and short fades) are best addressed in the modem, although early HF modems lacked the processing power to do this.

This chapter reviews the evolution of HF modem technology, followed by a discussion in some detail of the state-of-the-art modems and protocols used for military data communications in narrowband 3-kHz channels.

3.2 Data Waveforms

In simplest terms, the task of a modem is to convert a stream of user data bits into an analog signal (a stream of analog symbols) that can be passed through the audio passband of communicating radios, followed by conversion back into a stream of (hopefully identical) bits at the receiving station(s). Early modem designs employed frequency shift keying (FSK), in which incoming bits selected one of a set of tones to be sent over the air. More recent designs have applied *M*-ary phase shift keying (M-PSK) or *M*-ary quadrature amplitude modulation (M-QAM) to one or more audio tones (subcarriers) that lie within the audio passband of the radio. OFDM—discussed in Section 3.2.1.2—is an example of the multiple subcarrier approach. Before delving into the details of today's high-performance serial-tone waveforms, we first explore the options for HF modem design.

3.2.1 Design Space

With the steady advance of technology, continuous wave (CW) operation displaced spark gap transmitters. CW was, in turn, supplanted by increasingly sophisticated FSK waveforms. The first parallel-tone modem (a 16-tone kineplex waveform) was developed and tested in the late 1950s. Each of these relatively simple waveforms carried data adequately in benign HF channels but offered little robustness to the fading, multipath, and noise of more challenging conditions. The electronic technology of the day could not provide adequate error correction and interleaving within the modem due the computational complexity of even simple schemes.

By the late 1970s and early 1980s, however, research in communication theory and microelectronics began to yield practical results for improving the reliability of HF data communications. Two leading designs emerged: a serial-tone approach, which benefited from advancing computer technology to enable adaptive channel equalization, and a parallel-tone approach. Both approaches have been implemented in commercial products, resulting in a long-running debate over serial-tone versus multitone waveforms for various HF applications.

In this section, we present and evaluate a number of alternative approaches for sending data reliably over the HF channel, including serial- versus parallel-tone modulation, and various approaches to error correction, including both channel coding and interleaving.

We will address multipath propagation, fading, and impulsive noise, each of which presents significant challenges to HF waveform designers.

- *Multipath propagation* results in intersymbol interference (ISI): echoes of the transmitted symbols overlap, making error-free demodulation quite challenging, even with a strong signal.
- In a *fading channel*, the received signal level varies relative to the received noise level, resulting in a signal-to-noise ratio (SNR) that often drops too low to demodulate usable data.
- *Impulsive noise* also reduces the SNR when the noise level rises, thereby reducing the instantaneous SNR of the signal and again causing the demodulation process to fail.

3.2.1.1 Dealing with Fluctuating SNR

To cope with the effects of the fading and noise found in HF channels, waveform designers employ a combination of forward error correction (FEC) and interleaving. The mechanism of FEC is to introduce deliberate redundancy into a bit-stream (or symbol stream) that will aid the receiver in correcting errors introduced by the channel. FEC codes [1] commonly used on HF channels are Reed-Solomon (RS) codes, Golay codes, BCH codes, TCM codes, and convolutional codes. Note that turbo codes have also been investigated for use on HF, but have not been used in military standards due to the fact that most standards require that all waveforms be free of intellectual property rights (IPR), and turbo codes are patented.

In general, such FEC codes are most effective at correcting the errors caused by the channel if errors occur in short bursts. Unfortunately, the ionospheric channel can produce long strings of errors during fades and noise bursts. Such error bursts will cause most FEC codes to fail; in some cases the error "correction" process actually introduces additional errors.

Using an interleaver can help alleviate this problem. The operation of deinterleaving at the receiver can separate a burst of errors in the received data into widely separated errors in the data stream that is processed by the FEC. However, this requires that the length of the interleaver be substantially greater than the length of the fade or noise burst. Several interleaver architectures (block, convolutional, and helical) can be found in the literature [2, 3]. The benefits of an interleaver come at a cost in latency, however. For broadcast applications, there is little disadvantage in using a long interleaver, but for data applications that employ an automatic repeat request (ARQ) protocol, long interleaver latencies can greatly reduce system throughput.

3.2.1.2 Dealing with Intersymbol Interference

Multipath propagation results in a time-spreading at the receiver. The magnitude of this spreading varies with latitude, season, time of day, and so on; many ionospheric paths exhibit delay spreads on the order of a few milliseconds. This means that HF data signals will suffer ISI of a corresponding magnitude. For long symbols (i.e., longer than the delay spread) this overlap occurs only at symbol boundaries (Figure 3.1(a)). For higher signaling rates, however, the demodulator may have to deal with overlapped echoes of multiple symbols at every instant (Figure 3.1(b)).

This observation leads to two approaches to dealing with multipath-induced ISI:

- Use symbol times longer than expected multipath, and don't use the intervals containing ISI in demodulation. This limits the signaling speed, and therefore, the data rate achievable for each modulated subcarrier. If higher data rates are desired, multiple subcarriers must be employed. This approach is used in second-generation automatic link establishment (an 8-ary FSK waveform), and in OFDM and other multitone waveforms.
- Modulate a single subcarrier at a high symbol rate, and attempt to undo the multipath distortion at the receiver (e.g., by using an adaptive equalizer or a maximum likelihood sequence estimator (MLSE)). Such a waveform is

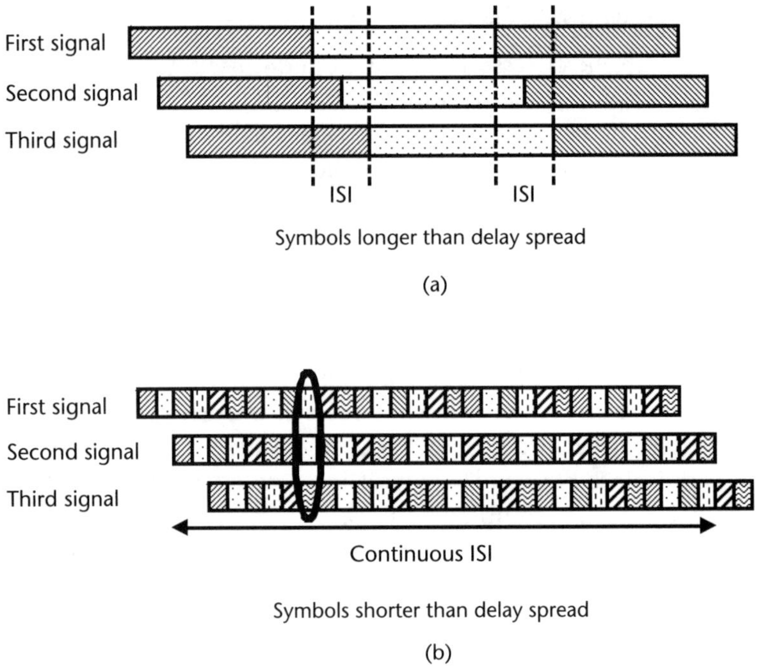

Figure 3.1 ISI arising from multipath propagation.

termed *single-tone* or *serial-tone*. Depending on the approach used for dealing with ISI, single-carrier waveforms may be much more computationally demanding to implement than multitone waveforms.

OFDM

OFDM refers to a class of multitone waveforms patented in 1970 [4]. Among the multitone waveforms in the literature [5, 6], OFDM is the most bandwidth-efficient and has the lowest computational complexity [7].

An OFDM signal is created by packing many tones within the audio passband of the radios and modulating each subcarrier independently (often using PSK or QAM). Crosstalk among the subcarriers is eliminated by making their signaling orthogonal: the tone spacing is an integral multiple of the frame rate.

With data carried on many subcarriers simultaneously, we can achieve a useful overall data rate even with a low frame rate on each tone. This allows the OFDM frame time to be longer than the delay spread of the channel. By including a cyclic prefix (guard time) in the frame, ISI (or more accurately, interframe interference) can be completely eliminated without the need for a complex equalizer [2] as long as the ISI length does not exceed the guard time. Research in OFDM without a guard time has been reported [8]. However, it seems likely that the adaptive frequency-domain equalizer that would be required for this approach is even more complex than the serial-tone equalizer described below.

The OFDM modulator and demodulator can be implemented efficiently using the fast Fourier transform (FFT), with each tone as one of the frequency bins of the

FFT. A more complete overview of the theory and implementation of OFDM can be found in Proakis [2].

An example of a parallel-tone waveform is the 39-tone waveform defined in U. S. MIL-STD-188-110B Appendix B [9]. Its frame length is 22.5 ms, with a guard time of 4.72 ms. Each of the 39 tones is modulated by DQPSK (differential 4-PSK) and four-bit Reed-Solomon codes are used for FEC. Since a differential modulation was used, no channel estimate is required to demodulate this waveform. However, if a coherent modulation was used instead of a differential modulation, a channel estimate and a single-tap equalizer (for each tone) would be required for proper demodulation.

One of the key limitations of OFDM waveforms—when used for data transmission on multipath fading channels—is frequency selective fading. This type of fading can cancel out or severely degrade the signal strength of many of the OFDM tones, producing an irreducible error rate. In the early 1990s, researchers combined some of the characteristics of code division multiple access (CDMA) and spread spectrum (SS) with OFDM in order to create a more robust modulation scheme that could survive frequency selective fading; thus OFDM-CDMA was born. Much of the original research focused on the uplink of cellular systems and how to best combine the benefits of OFDM, CDMA, and SS. OFDM was used to simplify the equalization process by the use of a guard time, which effectively reduces the equalizer to a single-tap complex multiplication (per tone) in the frequency domain. CDMA and SS were used to separate multiple asynchronous users operating in the same cellular channel communicating with a base station, and to create a more robust modulation scheme (SS in frequency domain can be viewed as *frequency diversity*). One additional benefit of this new system was that, by applying multiuser detection (MUD) techniques in the demodulation process (similar to CDMA), the capacity of the system could be increased.

On HF, OFDM-CDMA can be used in a completely different manner [10, 11]. Instead of many users sharing the same channel, data symbols can be treated as virtual users and spread across the frequency domain (instead of each data symbol modulating one of the available tones, as would be the case for OFDM). This spreading can effectively reduce the degradation caused by frequency selective fading on all the data bits, allowing for better performance on multipath fading channels. In addition, this approach yields a synchronous system, and there are no near/far[1] problems (as are typically encountered in CDMA cellular systems) because the virtual users are all sent at the same power level. Thus, when MUD techniques are applied at the receiver, the added computational complexity of asynchronous MUD and of the near/far problem can be disregarded. OFDM-CDMA offers some measurable performance benefits versus OFDM when uncoded waveforms are used [12]. As soon as interleaving and coding are added to both waveforms, OFDM performance is similar to OFDM-CDMA [12]. Although there may be some small added benefits to using OFDM-CDMA instead of OFDM on HF channels (slightly more robust to higher fade rates and narrowband interference), the additional receiver complexity may too high relative to the benefits obtained.

1. Near/far problems arise in CDMA cellular telephone systems when arriving signals have very different power levels. Received signal strength is much higher for users near the tower than for those far away.

Serial-Tone Waveforms

In a serial-tone waveform, the single subcarrier is modulated at a high symbol rate, limited by the channel bandwidth. For example, recent 3-kHz HF military waveforms employ an 1800-Hz tone modulated at 2400 symbols per second. This high symbol rate requires that the waveform be heavily filtered to fit into the allowed bandwidth. (The spectrum of this waveform before filtering has its first nulls at +/– 2400 Hz from the carrier.) At this symbol rate, the symbol length is 0.416 milliseconds (ms), so most ionospheric paths will introduce severe ISI. This ISI must be removed before a serial-tone demodulator can recover the transmitted data.

Several techniques have been developed for serial-tone waveforms to combat multipath [2, 13]:

- *Maximum-likelihood sequence estimator (MLSE);*
- *Adaptive equalization.*

An MLSE approach will achieve the best performance because it can use all of the signal energy arriving via the multiple paths, but its implementation complexity grows exponentially with the length of the channel impulse response and the modulation density (i.e., M-PSK, M-QAM). For example, if 64-QAM is used and the MLSE is to have L taps of multipath capability, the number of states of the MLSE would be 64^{L-1} (with 64 branches entering and leaving every state). This approach quickly exceeds the capabilities of today's processor technology. Reduced-state techniques have been developed that lower the complexity of MLSE but the effectiveness of these techniques has not been proven for the fading characteristics encountered on HF channels.

The alternative to MLSE, adaptive equalizers, provides reasonable complexity, but even the reduced computational complexity of an adaptive equalizer required custom hardware in the first serial-tone modems in the 1980s. It took another 10 years before off-the-shelf digital signal processing (DSP) technology could implement an adaptive equalizer for HF.

Another family of single-tone waveforms available to designers is continuous phase modulation (CPM) [14]. These waveforms offer some very attractive features, such as constant envelope and bandwidth efficiency. However, CPM is not widely used in HF applications; this is due mainly to the fact that it is a nonlinear modulation requiring an MLSE just to demodulate the waveform (and an even larger MLSE to demodulate the waveform in a multipath fading channel). It should be noted that some special cases of CPM do exist (such as GMSK) that allow the use of traditional adaptive equalizers. However, the general solution for CPM on multipath channels is MLSE. As noted above, the MLSE computational complexity is still too high for practical CPM waveform designs handling the same amount of multipath as current linear modulations (i.e., multipath spreads of 16 or more symbols).

3.2.1.3 Serial-Tone Versus OFDM Discussion

A number of myths and misconceptions about serial-tone and parallel-tone waveforms have arisen [15]:

- "Serial-tone waveforms are far superior to OFDM waveforms on HF." This conclusion arose because the MIL-STD-188-110B serial-tone waveform clearly outperformed the 39-tone waveform of the same late-1990s standard. However, research in the mid-1990s produced better OFDM waveforms that perform similarly to serial-tone waveforms.
- "OFDM waveforms are better than serial-tone waveforms for digital voice applications." Once again, differences were observed in equipment that used different technologies. In this case, the differences arose from the different FEC codes used for certain waveform standards (e.g., RS codes have different error statistics than convolutional codes) and not from any inherent property of the modulations.
- "OFDM is more bandwidth efficient and more power efficient than serial-tone waveforms." OFDM can, in theory, exploit the frequency-selective fading or nonuniform noise and interference encountered on an HF channel by adapting the data rate sent on each tone. For example, if a tone falls into a null of the frequency spectrum, it would be best not to transmit this tone at all, and instead increase the amount of data sent on the tones with the highest SNR. However, this approach has not been used on HF as it would be very complex to implement, requiring feedback from the demodulating station(s) to the transmitting station. It may not be possible to adapt quickly enough to keep up with the nonstationary HF channel.
- "OFDM is more robust to slowly fading channels because of its longer frame times." This is not a plausible claim. The fades observed on HF channels are longer than either the serial-tone symbols or the OFDM frames. Fading is handled for both types of waveforms by using an interleaver and FEC for the shorter fades, and an ARQ protocol for longer fades.

The remainder of this section contrasts serial- and parallel-tone modulation in terms of metrics related to waveform implementation and performance.

BER Performance

The bit error rate (BER) performance of a communication system is typically plotted as a function of signal-to-noise ratio (SNR) in a specified bandwidth (typically 3 kHz for HF). Such a BER plot is a straightforward approach to comparing the performance of waveforms (though measuring low BER points in a fading channel requires very long tests). Figure 3.2 shows the performance of the serial-tone and 39-tone waveforms (MIL-STD-188-110B, 2400 bps, long interleaving) on a standard channel (1-Hz Doppler spread and 2-ms delay spread). The serial-tone waveform performs significantly better on this channel.

Figure 3.3 compares the performance of a 2-PSK single-tone waveform to a more modern 2-PSK OFDM waveform both using coherent demodulation and the same FEC. The channel under test is the same as used in Figure 3.2. Even though the OFDM waveform uses perfect channel state information, while the single-carrier waveform must use a channel estimate, the single-carrier waveform outperforms the OFDM waveform by more than 1 dB. (Please see [16] for additional details.)

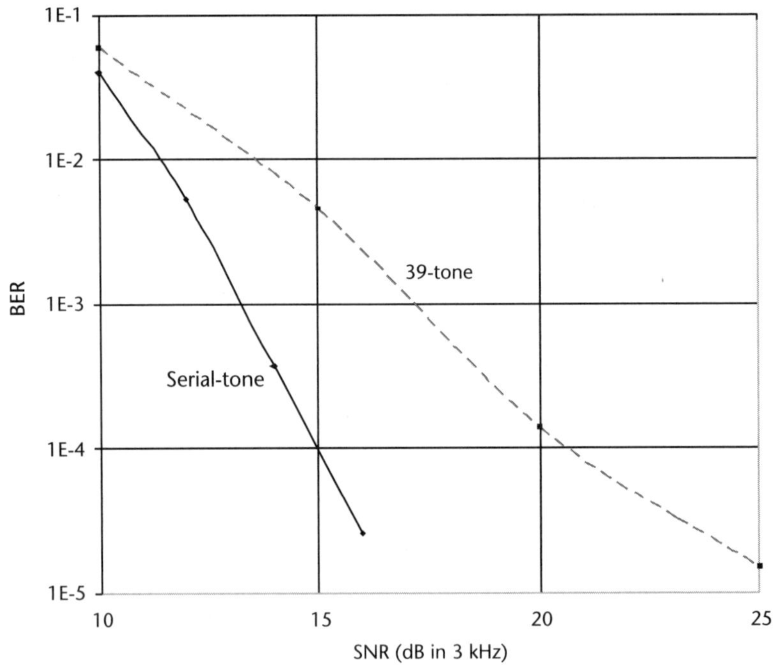

Figure 3.2 BER comparison of serial-tone and 39-tone waveforms. (After [15].)

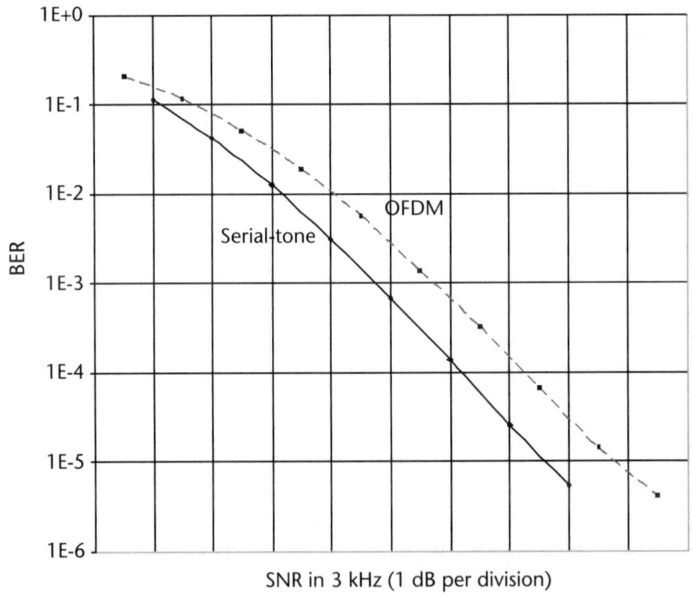

Figure 3.3 BER comparison of 2-PSK serial-tone and 2-PSK modern OFDM waveforms. (After [15].)

Power Efficiency and Bandwidth Efficiency

A serial-tone waveform will lose power and bandwidth efficiency due to the periodic insertion of known data to train the equalizer. An OFDM waveform will lose power and bandwidth efficiency in two ways:

- The insertion of a guard time. For the 39-tone waveform, about 1 dB was lost to accommodate a guard time of 4.72 ms (i.e., 4.72 out of 22.5 ms).
- The second source of loss arises when the waveform is provided with the ability to track the HF channel. For the 39-tone waveform, DQPSK was chosen as the modulation approach. Differential modulation removes the need for channel estimation without reducing throughput, but costs over 2 dB in SNR performance and limits the Doppler spread capability of the waveform. Another approach is to track the channel by transmitting known data as tones (usually referred to as *pilot tones*) and interpolating the known tones across time and frequency [17]. The losses for this approach depend on the ratio of known-to-unknown data inserted in the waveform to meet a desired Doppler spread and multipath capability.

Peak-to-Average Ratio

The peak-to-average ratio (PAR), or *crest factor*, of a waveform is defined as the peak envelope power divided by the average power. This ratio is meaningful because HF power amplifiers (PA) are usually peak power limited [18]. Thus, to avoid operating in the nonlinear region of a PA, the signal must be backed off by an amount proportional to the PAR.

The PAR of serial-tone waveforms arises from the filtering (analog and digital) required to constrain the waveform to a desired bandwidth and to the constellation used for modulating the tone.

For OFDM waveforms, the nonunity PAR is the result of the addition of the instantaneous amplitudes of all the tones in the time-domain, which yields a signal with a Gaussian-like amplitude distribution. The worst case PAR for a multitone waveform is N, where N is the number of tones. In practice, this worst case is seldom observed and most parallel-tone waveforms with $N > 20$ exhibit a 9- to 14-dB PAR, depending mainly on the number of tones and a small amount on the modulation and amplitude of each tone. The PAR can be further reduced by allowing the modulator to clip the waveform. Of course, this approach must be used with care since too much clipping will produce an irreducible error-rate: distortion of the tones in the frequency domain disrupts orthogonality and increases the noise floor. In recent years, other techniques to reduce the PAR of multitone waveforms have been developed, but they require either additional bandwidth [19] or additional processing on both the transmit and receive side [20].

Table 3.1 presents measurements of the average transmitted power for serial-tone and parallel-tone waveforms from MIL-STD-188-110B, measured using the 20W (peak) Harris RF-5800H tactical manpack radio with its internal modem. The received SNR was also recorded to show the effects of the transmit gain control (TGC) and automatic level control (ALC) functions in the transmitting radio. The 39-tone waveform was clipped for a PAR of about 6 dB.

Clearly, PAR has a large impact on the average transmitted power, and especially the receive SNR, when using a PA that is peak power limited. Note that this additional 2.2 dB of average transmit power (10W versus 6W) is seldom included in waveform design comparisons, but would provide a significant advantage to serial-tone waveforms when communicating on an HF link. Communication

Table 3.1 Average Transmitted Power and Receive SNR for MIL-STD-188-110B Waveforms

	Serial-Tone, 2400 bps Short Interleaver	*39-Tone, 2400 bps Short Interleaver*
Average power	10 watts	6 watts
RX SNR	30 dB	21 dB

engineers are strongly encouraged to include this difference at some point in the design comparisons so that the best on-air waveform is selected.

Narrowband Interference Excision

Narrowband interference (NBI)—such as a heterodyne tone or FSK modem tones—is not uncommon in HF communications. When NBI is not excised before reaching the demodulator, a serial-tone modem will suffer less from NBI than an OFDM modem. This is because NBI will overwhelm OFDM tones as the power of the interferer approaches the power per tone of the OFDM signal (total power divided by the number of tones). Serial-tone waveforms will not be significantly affected until the NBI power is large enough to degrade the SNR of the signal.

OFDM waveforms suffer from NBI in additional ways:

- If the frequency of NBI does not match one of the FFT frequency bins, NBI will spread into many frequency bins (instead of just one) because the implicit rectangular windowing of the FFT makes a tone look like a sinc function with slowly decreasing sidelobes.
- The FFT demodulation is very susceptible to discontinuities in the interfering signal (i.e., FSK frequency change, CW turning On/Off). A discontinuity will appear as rectangular windows of different lengths convolved in the frequency domain. Such discontinuities also affect the effectiveness of windowing techniques used to suppress the sidelobes of interferers.
- It is difficult for an OFDM demodulator to discriminate an interferer from data tones.

Research has been reported for improving the NBI excision capabilities of OFDM waveforms [21, 22], but most published results fail when NBI power is greater than or equal to the OFDM signal power (i.e., 0 dB interference-to-signal ratio).

On the other hand, serial-tone modems with NBI excision filtering have demonstrated impressive performances for a variety of interferers (CW, FSK, constant tone, etc.), especially at the lower data rates (i.e., 75, 150, 300, and 600 bps). The NBI performance specification of STANAG 4415 [23] requires that the 75-bps waveform handle interferers at power levels ranging from 20 to 40 dB higher than the desired signal.

Computational Complexity

The computational complexity of the serial-tone demodulation process is much higher than the demodulation process required for OFDM due to the need for an adaptive equalizer. This represents a clear disadvantage for serial-tone waveforms compared to OFDM. However, until the computational load required to implement a new, faster serial-tone waveform makes such a modem uneconomical, the advantages of the serial-tone approach make it more attractive. Consequently, serial-tone waveforms have been chosen for every United States government and NATO HF-waveform standard for many years.

3.2.2 PSK Serial-Tone Waveforms

In this section, we describe the initial PSK serial-tone waveforms that demonstrated the ability of modems with an adaptive equalizer to provide relatively fast, reliable data transmission over challenging skywave channels. In the mid-1980s, government-funded research in the United States produced a serial-tone waveform that served as the initial reference design for the NATO-authored STANAG 4285 HF [24] waveform. Minor modifications were made to this initial design to satisfy the parties involved in the development of 4285, and this STANAG was finally ratified in 1989. A second round of government funding (done to enhance the original waveform design) began in the United States in parallel with the STANAG efforts and led to the waveform defined in US MIL-STD-188-110A [25], which was officially published in 1991. These waveforms differ somewhat in details, but offer similar capabilities, with coded data rates of up to 2400 bps.

3.2.2.1 STANAG 4285

STANAG 4285 defined a simple serial-tone waveform that showed some of the performance advantages achievable with this class of waveform. Phase-shift keying of an audio subcarrier located at 1800 Hz (roughly in the center of the nominal 3-kHz passband) formed the basic modulation format of the STANAG 4285 waveform. The 2.4-kBaud symbol rate, coupled with appropriate pulse-shaping filters, permitted the modulation to be contained within the allowable 3-kHz allocations. The main body of STANAG 4285 defined three data rates of 3600, 2400, and 1200 bps, achieved with 8PSK, QPSK and BPSK modulations, respectively. Although not part of the main body of the STANAG, Annex E to the STANAG defined the constraint-length-7 convolutional code that normally provides the FEC coding for STANAG 4285 implementations. The coding scheme defined a puncturing of the rate-½ base code to map the 3600-bps uncoded rate to 2400 bps. The rate-½ code was used without puncturing to map the QPSK 2400-bps rate to 1200 bps and the BPSK 1200-bps rate to 600 bps. Lower rates of 300, 150, and 75 bps were achieved by repetition coding of the BPSK waveform. On-air, the data portion of the STANAG 4285 waveform appears to be composed of 8PSK symbols, regardless of the modulation selected. This is achieved by scrambling the modulated symbols with a pseudo-random sequence to ensure that if there are long constant inputs (a string of zero bits, for example), then the modulated signal will not degenerate to a tone. As the scrambling sequence is known to both the transmitter and receiver, the receiver is able to apply the correct BPSK, QPSK, or 8PSK decision boundaries in the demodulation process.

The basic signaling structure of the STANAG 4285 waveform is shown in Figure 3.4. The waveform comprises repeated 256-symbol segments. Within each 256-symbol segment, the waveform employs an 80-symbol preamble, followed by alternating blocks of 32 data symbols and 16 known training symbols. As shown in Figure 3.4, there are four blocks of data symbols and three blocks of known training symbols within each 256-symbol segment. The final data block of each segment is followed by the 80 known symbols of the next preamble.

Preamble Detection

The preamble of the STANAG 4285 waveform is designed for signal detection, frequency offset correction, and timing recovery. It uses the BPSK modulation of a repeated 31-bit m-sequence to fill out the 80 symbols of the preamble. A simple correlation-based receiver is sufficient to detect the signal in a benign channel. However, more sophisticated schemes are required to meet the 75-Hz frequency offset and 3.5 Hz/s Doppler sweep requirements of the standard.

The short length of the STANAG 4285 preamble can be a significant disadvantage in fading HF channels. To function optimally, the receiver must detect the presence of the signal and recover signal timing from a correlation against the *first* 80-symbol preamble of a transmission. For the lower rates (where the FEC coding provides significant redundancy) the detector may recover from missing the initial preamble by synchronizing on a subsequent 80-symbol preamble. In some cases, where sufficient redundancy exists, this will allow the FEC to recover the message without error. However, in many cases, the short preamble can prove problematic.

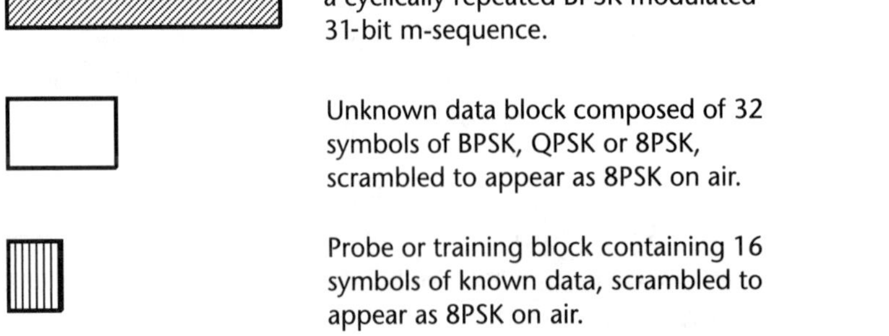

Figure 3.4 STANAG 4285 signal structure.

Serial-Tone Demodulation

The alternating blocks of data and training symbols within the STANAG 4285 waveform are designed to allow for demodulation using sophisticated equalization techniques. Annex C of the standard describes a conventional decision feedback equalizer and shows performance results obtained with that equalizer on simulated HF channels. Annex D, on the other hand, describes a more sophisticated equalization technique.

The most significant aspect of the equalizer described in Annex D is that it breaks the equalization problem into two distinct component processes. The first process is the estimation of the channel impulse response, and the second process is the detection of the data based on the received signal and the estimated impulse response. This general approach can be extended with equal success to equalization formulations quite different than those shown in Annex D.

The objective of the equalization process is to determine what symbols were transmitted based on the observed receive signal. In simple equalizers, such as the LMS-DFE [2, 26], the approach uses a feed-forward and a feedback filter, the coefficients of which are directly adapted based on the observed difference between the output of the filter and the decision made. In more sophisticated equalizers, the equalization procedure is divided into two separate tasks. The first is the estimation of the channel impulse response and the second is the estimation of the data symbols based on the received signal and the estimated channel impulse response. For example, the DFE approach described above can be modified to compute the feed-forward and feed-back filter taps based on the channel estimate [2]. The DFE is then used to process the signal and make decisions on received symbol values over the period for which the channel estimate is valid. Receive processing continues with a new channel estimate, computation of new feed-forward and feedback filter tap coefficients and subsequent detection of the next set of symbols.

The preferred approach is to include sufficient known symbols in the transmitted waveform so that the channel impulse response can be estimated directly from the known symbols. Alternatively, the channel estimate can be formed from past decisions, but this approach provides decidedly poorer channel estimates when errors are made in detection.

A channel estimate can be computed from the preamble and it can be maintained by a least-mean-squares (LMS) channel update routine. The well-known LMS update is a recursive procedure for updating an estimate, derived from a noisy gradient adaptive solution. Specifically, the recursion for updating the channel estimate, $\mathbf{f}_k$, is

$$\mathbf{f}_{k+1} = \mathbf{f}_k - 2\mu e_k \tilde{\mathbf{s}}_k^*$$

where μ is a step-size parameter, e_k is the error between the estimated channel output and the actual channel output, and $\tilde{\mathbf{s}}_k$ is the vector of training symbols and detected symbols in the channel estimator at time k. The usual manner for employing this technique in conjunction with a decision device is to run the decision device first, then update the channel estimate, then rerun the decision device and continue to iterate in this fashion either until the algorithm converges or—in a real-time system—a processing limit is reached.

The number of known symbols needed to compute a channel estimate must span at least twice the channel impulse response length. For most older serial-tone waveforms, the block of known symbols prior to (or following) the block of symbols to be demodulated is not long enough to be used to directly compute a channel estimate. It is in situations like this where the LMS update recursion just described will normally be used.

The length of the known (training) symbol sequence can impose a limit on the length of the channel impulse response that can be tolerated. Some detection strategies employ mathematical models that assume that the correlation matrix is Toeplitz. To satisfy this condition, the energy from the last unknown symbol of the data block must be received prior to the end of the training symbols that follow it. When this condition is not met, the matrix is not Toeplitz and other approaches must be employed.

The symbols detected by the equalizer are then mapped to bits. STANAG 4285 uses a Gray-coded modulation scheme. This is done to ensure that when a symbol, which represents 3 bits of information, is erroneously mapped to its nearest neighbor, only a single bit error is produced at the output of the receiver. This simple detection is sufficient in the rare case when no FEC is employed, but additional information of use to the convolutional error correction code is available from the detection. Soft decision information—which indicates the quality of each of the bits provided to the FEC—provides a significant performance advantage over coding schemes based on hard decision decoding. The ability of convolutional codes to effectively use soft decision information yields an advantage over other coding techniques (such as Reed-Solomon), which might otherwise provide better performance in a burst error environment [27]. The rate-½, constraint length-7 convolutional code employed by the STANAG 4285 waveform achieves good performance, but does require a flush at the end of the transmission.

The performance of the FEC can be enhanced by employing an interleaver to break up error bursts induced by the channel. At HF (because the duration of the fades can be quite significant), the interleavers must be much longer than those that are found in most other data communications signals. For STANAG 4285, a convolutional interleaver with two depths (0.8 s and 10.24 s) was chosen. This interleaver has the structure shown in Figure 3.5.

In contrast to more common block interleavers, the convolutional interleaver begins transmission of the data over the air immediately. At the same time, the initial fill (typically zeroes) of the convolutional interleaver is interspersed with the encoded data until the interleaving depth is reached and all data going over the air is encoded data. This interleaver has some interesting properties and provides significantly better interleaving depth than a block interleaver with the same end-to-end delay [28]. As a result of the choice of interleaver structure, the STANAG 4285 waveform defines a start of message (SOM) sequence to be sent as part of the data transmission. As with most waveforms, an end of message (EOM) is also included to indicate when the transmission is to be terminated. Note that convolutional interleavers must be flushed after the data to be transmitted has been sent (i.e., zeroes inserted at end of the user data).

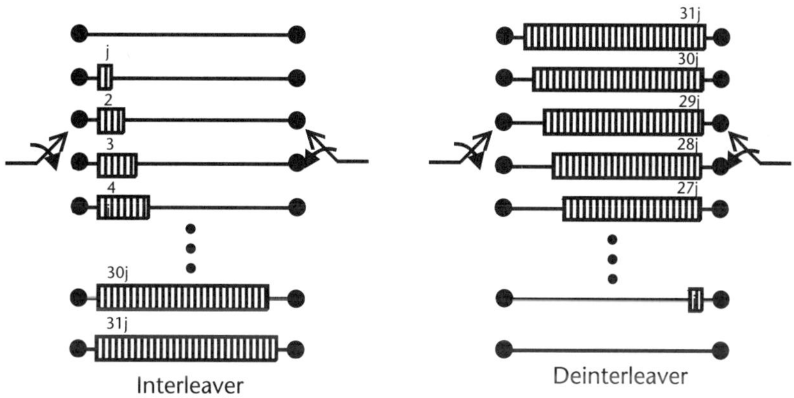

Figure 3.5 STANAG 4285 convolutional interleaver. (Adapted from Annex E of STANAG 4285.)

3.2.2.2 MIL-STD-188-110A Serial Tone

The serial-tone waveform defined in MIL-STD-188-110A shares many features with the STANAG 4285 waveform. Both waveforms use an underlying PSK modulation, with an audio subcarrier at 1800 Hz and a baud rate of 2400 Hz. Similar mappings of input bits to modulation symbols are used; in both cases, the data symbols are scrambled to avoid the generation of a tone in response to long repeated strings (of zeroes, for example) in the data. In each case, alternating blocks of unknown data symbols and known training symbols are used to facilitate the employment of sophisticated equalization techniques. In both cases, the FEC coding scheme is based on the same constraint-length-7 convolutional code. However, the MIL-STD-188-110A serial-tone waveform embodies several features not found in the STANAG 4285 waveform.

The MIL-STD serial-tone waveform uses a convolutional code very much like the Annex E convolutional code defined in STANAG 4285. The interleaving strategy, however, is quite different. With the MIL-STD serial-tone waveform, the length of the preamble is matched to the length of the block interleaver used. This design choice allows the transmit interleaver buffer to be filled while the preamble is being transmitted. With block interleavers of 0.6 s and 4.8 s, the preambles are of significant duration, providing substantially improved detection probabilities in poor signal environments compared to STANAG 4285. An additional benefit to the longer preamble is the ability to embed a few bits of robustly encoded information of use to the receiver. In contrast to STANAG 4285 (where the transmitter and receiver must agree in advance on the data rate and interleaver settings to be used), the 110A serial-tone waveform provides an autobaud feature where the receiver is able to determine the data rate and interleaver settings from the received waveform without needing to know them in advance.

The designers of the STANAG and MIL-STD waveforms also chose slightly different approaches to FEC coding and overall waveform design. With the STANAG waveform, the ratio between data and known symbols is always unity over segments beginning with a preamble and ending with the last data block before the next preamble. The difference in the design of the MIL-STD-188-110A serial-tone waveform is shown in Table 3.2 and Figure 3.6. For the two highest data rates—4800 (uncoded) and 2400 bps—the serial-tone waveform uses an 8PSK modulation with repeating blocks of 32 data symbols and 16 training symbols in the same fashion as the 4285 waveform. For rates between 150 and 1200 bps, a different block structure with 20 data symbols followed by 20 training symbols is employed. This change in block structure increases the waveform's ability to tolerate delay spread by increasing the number of training symbols from 16 to 20. This also modestly increases its ability to cope with Doppler spread by reducing the number of symbols in each data + training block from 48 to 40. On the other hand, the ability of a receiver to detect the absence of signal—or to recover from a

Table 3.2 MIL-STD-188-110A Waveform Summary

Data Rate	*Modulation*	*FEC Code Rate*	*Known Symbols*	*Unknown Symbols*
75	4-PSK x 32 symbols	1/2	none	all
150	2-PSK	1/8 (Rate 1/2 repeated 4 times)	20	20
300	2-PSK	1/4 (Rate 1/2 repeated 2 times)	20	20
600	2-PSK	1/2	20	20
1200	4-PSK	1/2	20	20
2400	8-PSK	1/2	16	32
4800	8-PSK	None	16	32

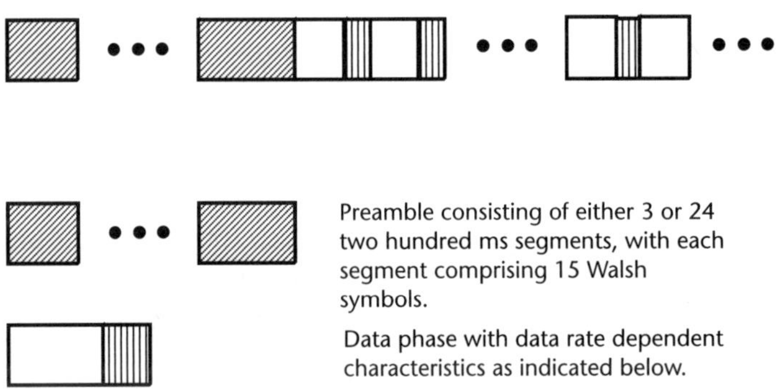

Data Rates	**Data Symbols**	**Probe Symbols**	**Modulation**
4800, 2400	32	16	8PSK
1200	20	20	QPSK
600-150	20	20	BPSK
75	-	-	Walsh

Figure 3.6 Serial-tone waveform structure.

prolonged period of poor signal—is somewhat worse than with a STANAG 4285 waveform, where the periodic reinsertion of large blocks of 80 symbols of known data provides better opportunities for assessing signal quality.

The different approach to waveform structure results in a different FEC coding scheme as well. Because there is no equivalent of the 80-symbol preamble reinsertion found with the 4285 waveform, when the waveform uses blocks of 32 data symbols and 16 training symbols, the proportion of data symbols is much higher than with the 4285 waveform (which has the same data and training block sizes). The term waveform efficiency is defined as the ratio of data symbols to total symbols transmitted during the portion of the waveform after the initial preamble. The STANAG 4285 waveform has a data efficiency of 50% for all data rates, while the MIL-STD-188-110A serial-tone waveform has a waveform efficiency of 66.7% for the highest rates (2400 and 4800 bps) and a waveform efficiency of 50% for rates between 150 and 1200 bps. As a result of the higher waveform efficiency for the 8PSK mode, the MIL-STD serial-tone waveform achieves a data rate of 2400 bps without having to puncture the convolutional code, as is necessary for the 2400-bps rate of the 4285 waveform.

The final significant difference between STANAG 4285 and the MIL-STD-188-110A serial-tone waveform is the provision of an entirely different modulation scheme to achieve the lowest (75-bps) rate. This lowest data rate waveform shares the same modulation scheme used in the preamble: a Walsh modulation[2] spread over many PSK symbols or chips. This 75-bps waveform allows for robust demodulation under conditions that are far worse than the repetition coded BPSK 75-bps waveform found in STANAG 4285.

MIL-STD Preamble

The MIL-STD serial-tone preamble uses repeated Walsh frames to allow synchronization at low signal-to-noise ratios and in high delay- and Doppler-spread channels. The duration of the serial-tone preamble is matched to the interleaver selected. For the long interleaver setting, the duration of the preamble is 4.8 s, which is the time required to fill the long (4.8 s) interleaver. For short or no interleaving, the preamble length is reduced to 0.6 s, which matches the time required to fill the short interleaver. With this selection—when used with a synchronous serial interface—the time that would otherwise be wasted while the interleaver was being filled is used to send the preamble.

The preamble is composed of distinct segments, with each segment being 0.2 s in duration. These segments comprise fifteen distinct 3-bit Walsh symbols (see Figure 3.7). Each Walsh symbol contains 32 chips; each chip is chosen from the 8PSK alphabet at the chip rate of 2400 Hz. The segment is structured as follows:

- The first nine symbols are fixed and known.
- The next two symbols encode data rate and interleaving settings.
- The following three symbols encode a countdown.
- The last symbol of 32 chips is again fixed and known.

2. In Walsh modulation, for each n-bit FEC-coded symbol to be sent, one of a set of 2^n orthogonal multisymbol sequences is transmitted.

0	1	3	0	1	3	1	2	0	D1	D2	C1	C2	C3	0

Figure 3.7 Preamble Walsh symbols.

For the short or no-interleaver selections, three preamble segments are transmitted. For the long interleaver selection, 24 preamble segments are transmitted.

Preamble Reception

One shortcoming of the MIL-STD serial-tone waveform definition is that the transmitted signal does not distinguish between no-interleaver and short interleaver. As a result, the receiver is able to uniquely distinguish the long interleaver from the other two possibilities (short interleaver or no interleaver) but is not able to distinguish short- from no-interleaver from the codes in the preamble. As a result, the transmitter and receiver must agree in advance whether the "short interleaver code" will represent no interleaver or short interleaver.

Data Phase

Once the preamble has been detected, if the data rate is 150 bps or higher, MIL-STD data detection proceeds in a manner equivalent to that described above for data detection with the 4285 waveform. For the previously discussed equalization technique, an estimate of the channel impulse response is formed using the received preamble. A trial detection of the first data block may be made using this estimate. Outputs from this trial detection can be used to improve the channel impulse response estimate, and a revised detection can be made. This process may be iterated multiple times to improve performance in fading channels, an important feature in channels that change rapidly.

To a first order, the ability of the waveform to tolerate Doppler spread is proportional to the inverse of the time between successive channel probes. A useful way of thinking about this is to look at the channel estimation as a sampling of the real channel as it changes with time. Clearly, when a time-varying process is sampled more often, the effects of the variation with time can be better approximated. As a result, neglecting the advantage that the MIL-STD serial-tone waveform has in FEC code rate, it would be expected that the highest rates of the MIL-STD and STANAG waveforms would show similar performance because they both employ the same underlying data and training block sizes.

With an equalizer that assumes a Toeplitz correlation matrix, the ability of the detector to tolerate delay spread is explicitly tied to the length of the training symbol segment. For other equalizer structures, performance declines, but the receiver does not fail catastrophically when the delay spread exceeds the length of the training symbol segment. In either case, this dependence on the length of the training symbol segment leads to an advantage for the 150- to 1200-bps rates of the MIL-STD serial-tone waveform, where the 20 data symbol/20 training symbol structure is used. The additional four symbols of known data allow this waveform to tolerate about 1.7 ms more delay spread than waveforms based on block structures with 16 training symbols.

Walsh Modulation for the 75-bps Data Rate

Figure 3.8 is a block diagram of the 75-bps data phase transmit processing. During the data phase, user data bits are encoded by a rate-1/2, constraint-length-7 convolutional encoder. The output bits of the encoder are then loaded into a block interleaver structure. While one interleaver block is being filled, an alternate block is being emptied to generate the transmit waveform. Two bits at a time are removed from this interleaver structure, forming a two-bit modulation word.

Past this point in the waveform generation, the modulation differs significantly from the MIL-STD-188-110A higher bit-rate modes and from all the bit-rate modes of STANAG 4285. Instead of directly generating a PSK signal from the two modulation bits, these bits are used to select one of four orthogonal Walsh functions [29] listed in Table 3.3.

The 4-ary orthogonal modulation is performed by taking the four-element sequence displayed in Table 3.3 associated with the two-bit modulation word and repeating this sequence eight times, resulting in a 32-element vector. This sequence of 32 symbols is then scrambled in the same manner as the individual symbols at the higher data rates. After the orthogonal Walsh modulation is complete, the 32-symbol 8-PSK sequence is low pass filtered and used to modulate an 1800-Hz subcarrier at 2400 symbols (chips)/s.

The above processing is repeated each frame time, with the 32 element scrambling sequence repeating after every five frames. Additionally, at the end of each interleaver block, the transmit waveform is slightly altered to use a higher order set of the Walsh functions, thereby identifying the interleaver block boundary.

At the termination of the user data, a 32-bit EOM sequence is passed through the encoder and interleaver. This EOM is followed by 144 flush bits (all set to 0) to

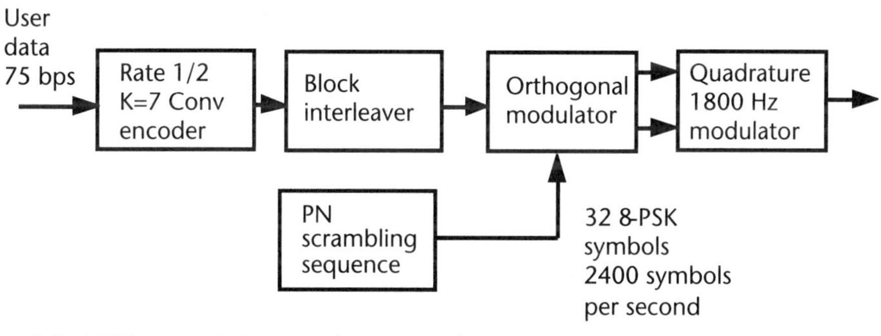

Figure 3.8 Walsh transmit data waveform generation.

Table 3.3 Orthogonal Walsh Function Elements

Mod. Word	*4-ary Walsh Function Elements*			
00	0	0	0	0
01	0	4	0	4
10	0	0	4	4
11	0	4	4	0

flush the encoder. After the 144 flush bits are input, additional 0 bits are encoded until the final interleaver block is filled completely. The transmission of this interleaver block terminates the 188-110A 75-bps transmission.

The optimal receiver structure for a spread signal (which is wideband relative to the coherence bandwidth of the channel process) is known to be the RAKE or correlation receiver; this receiver structure has been covered extensively in the literature (e.g., [2]). This structure relies on an estimate of the impulse response of the channel, which is used to combine the multipath components resolved by the RAKE receiver structure. Calculation of the optimal decision variables depends on an accurate estimation of the channel impulse response.

Real-time modem implementations must also deal with the real world issues of frequency offset estimation and time tracking. Frequency offset can be caused by mistuned or inaccurate radio references, or a high-speed moving platform. Differences in the sampling clocks or time references between two modem platforms always cause a time drift of some magnitude between the transmitted waveform and the receive signal processing. To combat this degradation, the receive modem must track the offsets to implement corrections to its time base.

3.2.3 MIL-STD-188-110B and STANAG 4539

In the mid-1990s, researchers at the Communications Research Centre in Canada were conducting a field trial to characterize improvements that could be obtained using polarization diversity, both for transmit and receive operation [30]. The test signal for the trial was to be transmitted alternately on vertically and horizontally polarized antennas, while the signal was received and recorded on a distant pair of colocated vertically and horizontally polarized antennas. With this setup, it was possible to look at all combinations of transmit and receive polarization and, on receive, to directly compare receptions from either single antenna with those that combined the signal from both antennas in a diversity arrangement.

The test signal consisted of sequential transmissions of various data rates achieved with single-tone modulation, combined with FEC based on convolutional codes. The increase in available processing power (compared to 1980s DSP technology) meant that modified equalization techniques, capable of effectively using more efficient waveforms, were practical. In an attempt to characterize the diversity gain performance as fully as possible over a wide range of conditions, the PSK modulations used in the trial mapped to data rates from 75 bps BPSK with repetition coding to 4800 bps using a punctured 8PSK modulation. Interestingly, signals based on QAM modulations were also included, which provided a series of data rates up to 9600 bps as part of the test signal.

The design intent of the test signal was to have a selection of signals which, at the low data rate end, were more robust than would ever be needed on the link, and on the high end, were more complex than could be supported by a skywave link. However, the results of those trials led to the recognition that the higher order QAM constellations were practical not only for use on surface-wave channels, or in conjunction with diversity combining, but in many cases worked well over skywave channels.

During the same time frame, the early drafts of NATO STANAG 5066 were being written. This new profile for HF data communications clearly stood to benefit from the provision of a new standardized waveform to achieve data rates greater than those offered by the existing STANAG 4285 and MIL-STD-188-110A serial-tone waveforms. This led to the inclusion of Annex G in the draft STANAG 5066, which defined a waveform providing data rates of 3200, 4800, 6400, and 9600 bps, using Q PSK, 8 PSK, 16 QAM, and 64 QAM, respectively. The choice of data rates above 2400 bps was a decision to avoid conflict and interoperability concerns associated with providing the 2400 bps rate already standardized in other waveforms.

Sadly, in international standards efforts, things rarely go smoothly. The NATO groups responsible for standardization felt that the new waveform did not belong as an Annex to STANAG 5066, which was categorized as a protocol standard. Instead, they preferred a separate STANAG to define the waveform. Competition to define the new NATO-standard high-speed waveform ensued between groups in France, Germany, and a cooperative effort by developers in Canada and the United States. It was this latter cooperative effort that spawned Appendix C of MIL-STD-188-110B. The new waveform used the constellations and much of the waveform structure developed for STANAG 5066 Annex G and added further enhancements: a 32QAM constellation providing an 8 kbps data rate; a tail-biting convolutional code; a block interleaver with six interleaving depths; and a preamble that supported a full autobaud feature.

The competition held by the NATO working group pitted three competing designs against one another, with scientists from the Defence Evaluation and Research Agency (DERA) in the United Kingdom conducting the testing on behalf of NATO. The entry from France was based on an OFDM modulation scheme utilizing a cyclic prefix and pilot tones in conjunction with a turbo-code FEC scheme. The entry from Germany was based on a PSK/QAM architecture, with a symbol rate significantly higher than the 2400 baud used by previous serial-tone waveforms. The higher symbol rate resulted in some degradation in performance when the waveform was constrained by radio filters to fit within the usual 3-kHz channel allocation. The final entry in the competition was an implementation of MIL-STD-188-110B Appendix C developed by the Harris Corporation of Rochester, New York and Melbourne, Florida.

The results of the competition were unambiguous, with the new MIL-STD waveform emerging as the clear winner. The end result of this standardization process was that the two standards (STANAG 4539 and MIL-STD-188-110B Appendix C) adopted the same waveform. The performance specifications in the STANAG were slightly different, as they required the use of radio filters in the simulation environment, and the required performance targets in the STANAG were somewhat more difficult to meet, to ensure that STANAG modems reach the performance levels not achievable by the competing waveforms.

3.2.3.1 Overview and Waveform Structure

The characteristics of the STANAG 4539 and MIL-STD-188-110B Appendix C waveform are summarized in Table 3.4.

The waveform structure is shown in Figure 3.9. To facilitate transmit ALC and receive AGC functions, the transmitter may optionally send up to 7 repetitions of the complex conjugate of the first 184 symbols of the initial preamble. These

Table 3.4 MIL-STD-188-110B Appendix C Waveform Summary

Data Rate	*Modulation*	*FEC Code Rate*
3200	4-PSK	3/4
4800	8-PSK	3/4
6400	16-QAM	3/4
8000	32-QAM	3/4
9600	64-QAM	3/4
12,800	64-QAM	None

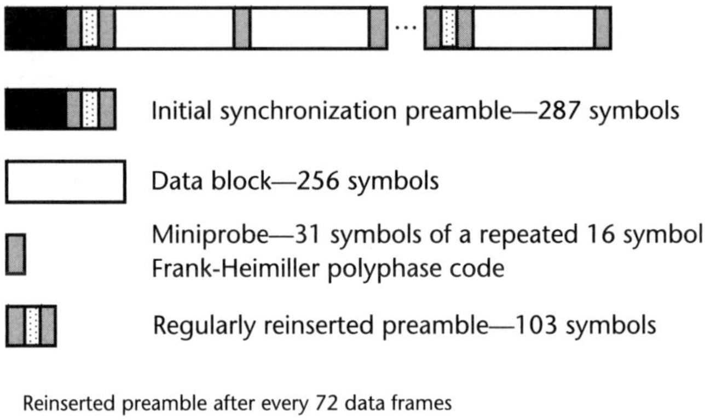

Figure 3.9 Appendix C waveform structure.

first 184 symbols of the initial preamble are a pseudo-random BPSK modulated sequence, chosen for good correlation properties to aid signal detection. This is followed by a 103-symbol segment comprised of two 32-symbol segments, each containing a cyclic repetition of a length 16 Frank-Heimiller (FH) sequence, separated by three 13-symbol Barker sequences. The three Barker sequences are quadrature modulated to encode the data rate and interleaver settings. This allows the receiver to decode the transmission without a priori knowledge of the data rate and interleaver settings (the famous MIL-STD autobaud capability).

The initial preamble is followed by alternating blocks of 256 data symbols and 31 known miniprobe symbols, where the miniprobe symbols are constructed from the same cyclically extended 16-symbol FH sequence.

The designers of the Appendix C waveform thought that it would be useful to include a regularly reinserted preamble. This regularly reinserted preamble is the same as the final 103 symbols of the initial preamble, and allows acquisition of the signal if the initial preamble is missed and also simplifies timing correction for long transmissions. The regularly reinserted preamble is the reason for the 31-symbol miniprobe length.

A waveform made up of blocks with 256 data symbols and 32 known symbols has a waveform efficiency of 8/9. When coupled with a symbol rate of 2400 bps and a rate-3/4 FEC code, this leads to desirable data rates for synchronous interfaces for most modulations. For example, with 6 bits per symbol, it results in a

9600-bps user data rate. However, periodically reinserting a preamble will reduce the effective efficiency of the waveform and hence throw off the data rate calculation. In the Appendix C waveform, the 103 symbols of the reinserted preamble may be thought of as being composed of 31 symbols of the FH sequence from the data block immediately preceding the reinserted preamble, plus 72 additional symbols. The efficiency penalty for each of these additional symbols has been made up by using 31-symbol miniprobes (instead of 32 symbols) in each of the 72 blocks between reinserted preambles.

The FH-based miniprobe sequence that is transmitted can be one of two phases, positive (+) or inverted (–). Miniprobes are modulated, positive or inverted, with the autobaud information over the 72 blocks between reinserted preambles. This allows the receiver to infer the data rate and interleaver setting from only the miniprobes, without necessarily waiting for the reinserted preamble. In practice, this feature does not work effectively in real HF channels. In most cases, if the initial preamble is missed, signal acquisition will coincide with the reinserted preamble.

3.2.3.2 Modulation: PSK and QAM

The most innovative feature of the Appendix C waveform, when compared to previous HF waveforms, is its use of quadrature amplitude modulation (QAM) for data rates of 6400 bps and higher. 16-, 32- and 64-QAM constellations have been used. The choice of QAM constellations (rather than PSK constellations) is mandated by the need to improve bandwidth efficiency in order to achieve higher data rates, and by the increasingly poor signal space distance properties of PSK constellations beyond 8-PSK. The 16-QAM and 64-QAM constellations are shown in Figure 3.10. A notable feature of these constellations (relative to classical square QAM constellations) is the way points fill the area within a circle of constant radius, maximizing signal space distance, while minimizing peak to average ratios.

These constellations have been carefully designed and retain the desirable Gray-coding features of the square constellations. That is, for most of the constellation

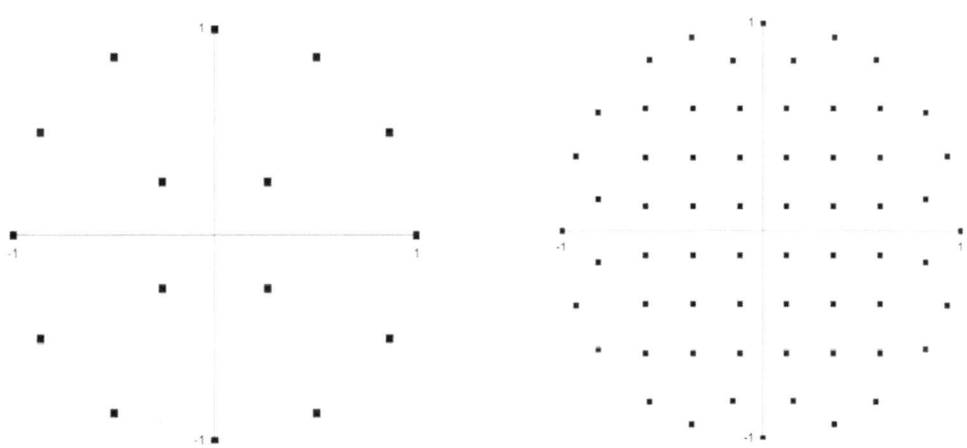

Figure 3.10 HF QAM constellations.

points, a decoding error that results in an incorrect selection of a nearest neighbor results in only a single bit error in the output.

3.2.3.3 Demodulation of Data Blocks

The data blocks of the Appendix C waveform consist of 256 symbols of data to be demodulated with cyclically extended 16-symbol FH sequences immediately before and immediately following the data block. The cyclic FH sequences have perfect correlation properties, and as such, provide good channel estimates before and after the data block. These channel estimates are based solely on known data and do not depend in any way upon decisions made by the receiver. This is different from the equalization approach normally used with previous generation waveforms where an LMS update procedure incorporates decisions made on data symbols to update the channel estimate.

The length of the FH sequence limits the length of the channel impulse response estimation to 16 symbols, or 6.7 ms, when this approach is used. This results in a delay spread capability that is in line with 110A and STANAG 4285 waveforms. The Doppler spread capability is somewhat reduced, however. While the provision of good channel estimates before and after the data block improves performance significantly, relative to having to use known data in the channel estimation process, the large interval between channel estimates results in somewhat poorer resistance to Doppler spread compared to the older, lower-data-rate waveforms.

In most cases, the increased waveform efficiency (8/9 for Appendix C versus 2/3 for 110A 2400 bps or STANAG 4285) results in better performance by the Appendix C variant because of the ability to use lower modulation complexity to achieve the same data rates. For example, the 3200-bps data rate in Appendix C provides better performance on most channels than the 2400-bps data rate in either MIL-STD-188-110A or STANAG 4285, as it uses a QPSK modulation instead of 8-PSK.

3.2.3.4 Forward Error Correction

The Appendix C waveform introduced a fully tail-biting convolutional code as part of the FEC scheme. Tail-biting improves efficiency significantly in packet-based systems, as it eliminates the need to flush the encoder at the end of a transmission (as was done previously). With the new tail-biting FEC, it became possible to precisely match the data to be transmitted to the duration of the time slot available, an important feature for optimizing ARQ or slotted transmissions

The FEC scheme has additional features that benefit packet-based transmissions. Like the 110A waveform, the Appendix C waveform uses a block interleaver. However, a much wider range of interleaving options is provided, with 6 choices ranging from an ultrashort interleaver of 0.12 s to very long interleaver of 8.64 s. The very long interleaver provides good performance in fading for broadcast transmissions, while the shorter interleaver options allow users to make trade-offs between latency and resistance to fading effects. In many applications, the cost of the long latency associated with the very long interleaver overshadows whatever gains result from the performance advantage in fading channels.

3.3 ARQ for HF Radio Data Links

The modem waveforms presented so far include sophisticated mechanisms for coping with the short-term variations of the HF channel. However, as noted in the introduction to this chapter, longer-term variations require adaptive features at higher layers of the protocol stack. Therefore we now discuss link-layer protocols that improve the reliability of data transmission over HF channels through the use of automatic frame error detection and retransmission.

3.3.1 Introduction to ARQ Protocols for HF Radio Links

At the data link layer, packets to be sent are segmented into *frames*, which are contiguous segments of a few hundred bytes. For error detection, a checksum or cyclic redundancy check (CRC) code is computed for each frame of data to be sent, and then appended to that frame. The receiver recomputes the error detection code from the bytes received. If the result doesn't match the code sent along with the data, the receiver requests retransmission.

Most HF data links operate in one direction at a time, so HF automatic repeat request (ARQ) protocols are cyclic: Station A sends one or more frames of user data (the forward transmission) to Station B, the link direction is reversed (Station A becomes the receiver and Station B the transmitter), Station A sends an acknowledgment (ACK), and the link is again turned around. This cycle is repeated until the entire user message or file is successfully sent.

The link turnaround time on an HF link can be much longer than for other wireless media. These delays, on the order of a second or two, arise from the need to employ relatively long interleavers on HF links. During each link turnaround, the sending modem continues to send for at least one interleaver time after the sending protocol has stopped delivering data to that modem; then the direction of the physical layer can be reversed. After the RF electronics have switched from transmitting to receiving and vice versa, the new sending station must send a modem preamble before its first interleaver of data reaches the channel. These long link turnarounds, and the quantization of payload transmission times into interleaver-size units, strongly influence the design of ARQ protocols for HF radio links.

In general, ARQ protocols fall into one of three classes:

- *Stop-and-wait protocols* are the simplest. A single frame of data is sent. The sender then waits for a positive or negative ACK to arrive within a bounded time. If a negative ACK arrives, or the timeout expires, that frame of data is re-sent. If a positive ACK is received, the next frame of data is sent. The receiver discards frames received with error(s), and delivers error-free frames to its higher-layer client. Stop-and-wait ARQ protocols can be inefficient on HF links because of the short forward transmission. First, the single data frame may not fill even one interleaver, so the remainder of that interleaver must be filled with throwaway bits. Second, the sum of the turnaround times may exceed the forward transmission time, in which case productive use of the channel falls below 50%.

- *Go-back-N protocols* send multiple data frames in each forward transmission before pausing for an ACK. This leads to increased efficiency over stop-and-wait protocols because the time lost to link turnarounds and receiving the ACK is amortized over all of the frames in the forward transmission. In go-back-N ARQ protocols, the ACK indicates the *first* frame from the preceding forward transmission that must be re-sent. All frames following that missing frame must also be re-sent. Although this can be inefficient, it means that the receiver is not required to store frames received out of order.
- *Selective repeat (also known as selective acknowledgment) protocols* are the most complex, but also the most efficient ARQ protocols for HF channels. As with go-back-N protocols, the forward transmission carries multiple frames, but the ACK indicates exactly which frames were received error-free. This eliminates the need to resend frames that had no errors at the receiver (seen in go-back-N protocols). However, the receiver must store the frames received out-of-order, along with the information needed to reassemble those frames into complete client data packets once all frames have been received error-free.

Figure 3.11 illustrates these three types of protocol in operation on a fading HF channel. In each example, the channel fades twice at the same points in time.

- In the *stop-and-wait example*, the first fade corrupts the first ACK frame. The transmitter timeout expires, and the transmitter resends frame 1. An ACK is received, so the transmitter sends frame 2, but that data frame is corrupted and the receiver sends no ACK. After the timeout expires, the transmitter resends frame 2. From that point onward the link suffers no losses.
- In the *go-back-N example*, frames 2 and 6 are corrupted, so only frame 1 is acknowledged. The transmitter therefore starts its second forward transmission with frame 2, followed by the succeeding contiguous frames.
- In the *selective-repeat example*, the same two frames (2 and 6) are corrupted, so the receiver acknowledges frames 1, 3, 4, 5, 7, and 8 as received error-free. The two missing frames are re-sent in the next forward transmission, followed by new frames starting with frame 9.

Note that the turnaround times were depicted as shorter than the frame time. If the turnaround times were instead longer than the frame time, the stop-and-wait protocol would have fallen far behind even the go-back-N protocol instead of roughly keeping pace.

3.3.2 FED-STD-1052

For interoperable data communication over HF channels, we must employ common technologies at each layer of the protocol stack (as well as agree on standard operating procedures among organizations that wish to interoperate). With the standardization of the serial-tone HF data modems in MIL-STD-188-110A, it quickly

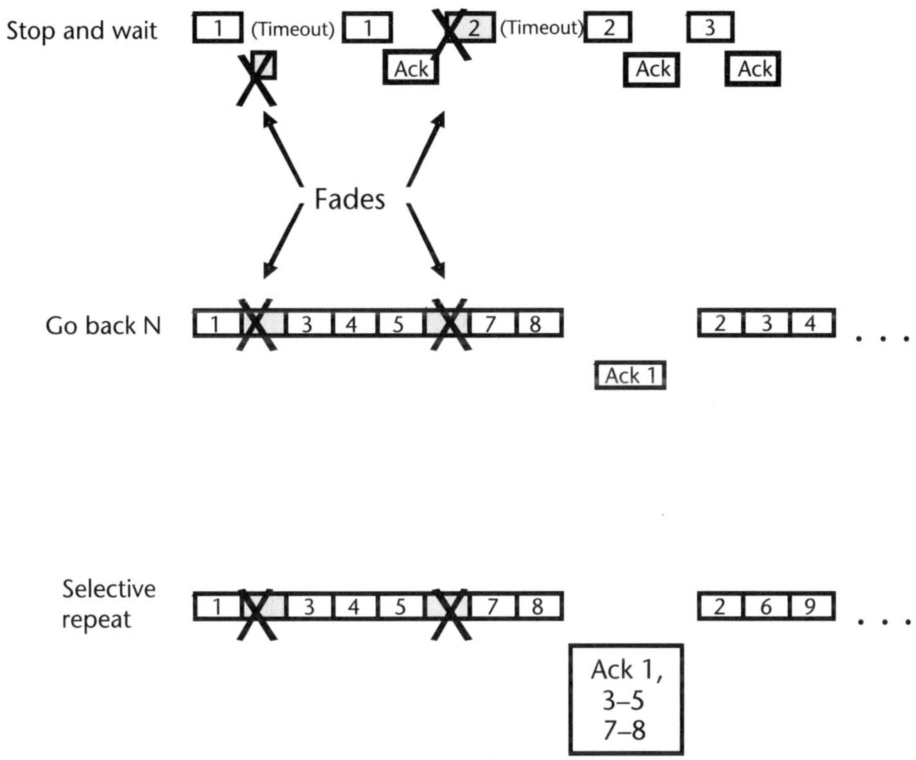

Figure 3.11 Illustration of ARQ protocols in operation.

became apparent that a standard data link protocol was also needed. In the United States, the federal government standardized and published a selective repeat ARQ protocol for HF radio in FED-STD-1052 in 1996.

The *1052 protocol* enjoyed some success for the rest of the decade, but suffered from a collision-avoidance timing problem. In general, wireless data systems cannot listen while transmitting. This is due to the large difference in power levels between the transmitter output and the signals arriving at the receiver from a distant node. This is especially true for HF radio links, in which transmitter power amplifiers operate at tens, hundreds, or thousands of watts. Thus, a station sending ACKs must wait to transmit until the station sending data has finished transmitting; otherwise its ACK may go unheard.

We run into a conundrum when the forward transmission fades for a long time at the receiver: how long must that receiver wait to ensure that its ACK will not collide with the faded but possibly ongoing forward transmission? In FED-STD-1052, the solution was to wait for the longest possible transmission to complete before sending an ACK, which resulted in significant delays on fading links. This is one reason that the FED-STD-1052 protocol was largely abandoned when NATO standardized a selective repeat ARQ protocol for HF links in STANAG 5066 that did not suffer from this problem.

3.3.3 STANAG 5066

A NATO project to develop a standard profile for maritime HF data communications evolved into a widely adopted specification for an HF subnetwork service used by many applications. The current version of this standard is STANAG 5066 Edition 2, promulgated in 2008 [31]. (Edition 3 has been prepared but not yet promulgated.)

A notional diagram of a STANAG 5066 application is shown in Figure 3.12.

- Users of the service are application programs, such as email clients, running on the client computers shown in the figure.
- Client computers are connected via a local area network (LAN) to a STANAG 5066 host computer. Application programs running on these client computers create socket connections to the 5066 host computer to send and receive packets and to control the subnetwork service.
- In military applications, the client application programs may be exchanging classified data, so a cryptographic device is shown between the host computer and the physical layer (modem and radio). This would not be needed for unclassified applications.
- The HF management unit shown on the left side of the figure provides automated adaptive services that establish and maintain HF links for use by the STANAG 5066 subnetwork service. The details of these services are not specified in STANAG 5066. We will discuss automatic link establishment (ALE), automatic link maintenance (ALM), and automatic channel selection (ACS) in Chapters 4 and 5.

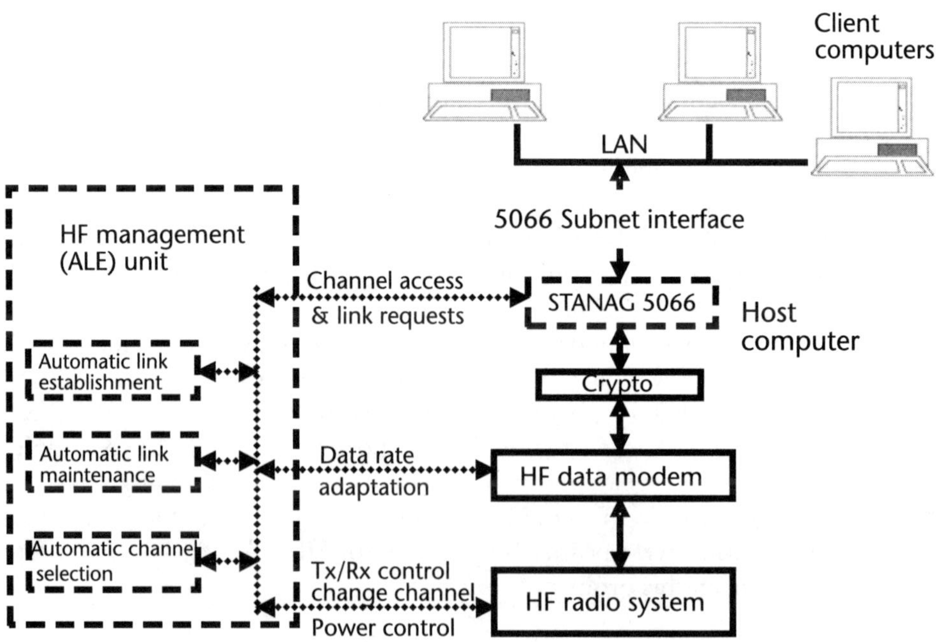

Figure 3.12 Notional application of STANAG 5066.

The layered structure of the STANAG 5066 subnetwork service is shown in Figure 3.13. The three sublayers in the figure that are not shaded are specified in STANAG 5066. These sublayers are briefly discussed in this section as well as a selection of standardized clients.

3.3.3.1 Data Transfer Sublayer

The data transfer sublayer (DTS) in Annex C to STANAG 5066 specifies a selective repeat ARQ protocol that has a number of attractive features:

- The header of every frame includes an *end-of-transmission field*. This 8-bit field specifies how much of the current transmission remains, in units of one-half second. This elegantly eliminates the end-of-transmission ambiguity that arises during an extended channel fade (noted above for the FED-STD-1052 protocol). If even a single header is received error-free, the receiver knows when it will be safe to send an ACK. Note that this field bounds the duration of STANAG 5066 transmissions at just over two minutes.
- A data-and-ACK frame type is specified: ACKs are carried in the frame header, and client data is carried in the body of the frame. In the simple ARQ cycle described earlier, one station sends data in a forward transmission, while the other is only permitted to send ACKs. By using data-and-ACK frames, every station in a STANAG 5066 link may send forward transmissions when it is its turn to transmit, appending ACKs for data recently received onto its data transmissions. This eliminates the overhead that would be needed to manage the direction of data flow on the client-to-client link if the protocol permitted only unidirectional data flow.
- A sliding window flow control mechanism is nicely integrated with the selective repeat acknowledgements; this reduces the number of selective-ACK frame numbers that must be sent. Frames are numbered using an 8-bit number that wraps around from 255 to 0 (i.e., it is incremented mod 256). The receive lower window edge (RX LWE) reported in the frame header identifies

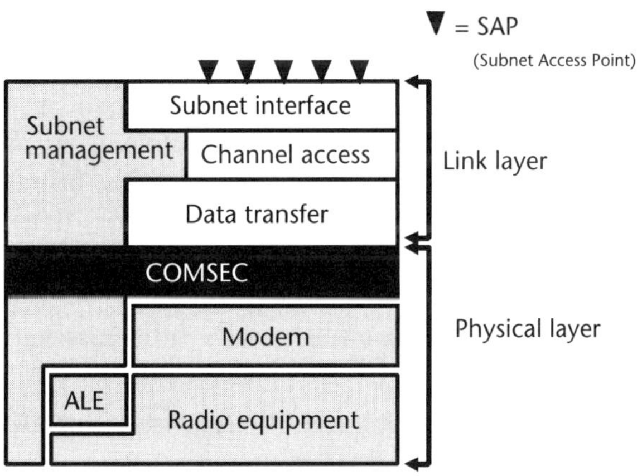

Figure 3.13 Layered structure of STANAG 5066.

the lowest-numbered frame (mod 256) that has not yet been received; this implicitly acknowledges all lower-numbered frames (mod 256), and also sets the lower boundary of a receiver flow control window of 128 frames that may be outstanding on the link. When the RX LWE in a newly received header is greater than the previously received RX LWE by more than 1 frame, all of the intervening frames are acknowledged without the need to individually list them.

Frame Size

The data section of a payload-carrying frame may be up to 1023 bytes in length. Its integrity is checked by a 32-bit CRC that is separate from the (16-bit) CRC on the frame header. The amount of data actually carried in each data frame is announced in the header of that frame. How should we determine the size of each fragment of the payload that will be carried in data frames? Efficiency favors larger fragments to dilute the overhead of the header and CRC on each frame. However, shorter frames are likely to be more robust in the presence of the high error rates sometimes experienced on HF channels. The standard recommends a default frame size of 200 bytes, a good compromise for many applications.

Data Rate Adaptation

A powerful adaptive feature of HF data link protocols, including both FED-STD-1052 and STANAG 5066, is the ability to adjust the data rate of the modems used to carry their frames over the HF channel. This will allow reliable operation over an SNR range of 30 dB or more by using a waveform whose robustness is suitable for the current SNR, and adjusting the waveform as the SNR varies on time scales commensurate with the ARQ protocol cycle time.

Practical implementation of data rate adaptation requires two things: (1) a channel metric that is accessible to the data link protocol, and (2) an algorithm for selecting a modem waveform using that metric.

For the channel metric, the most natural candidate would be the SNR. This is measured by many HF modems. However, in military applications, the 5066 host computer is separated from the modem by a cryptographic device, so communicating with the modem would require a bypass mechanism, which must be approved by the relevant security agency. Many 5066 implementations therefore use an alternative channel metric that is directly available—the frame error rate (FER).

The algorithm used in FED-STD-1052 for data rate adjustments was straightforward:

- If fewer than half of the frames in a transmission were received error-free, decrease the data rate.
- If all of the frames were received error-free, increase the data rate.
- Otherwise, make no change.

More recent research [32] determined that overall throughput is optimized by using FER thresholds that depend on the current data rate, as shown in Table 3.5. At the lower data rates that were in use when FED-STD-1052 was published, reducing the data rate at 50% FER was optimal. However, at the higher MIL-STD-188-110B data rates, one step downward in rates does not halve the rate, so we are better off reducing the rate at lower FER thresholds. Trinder [32] also found that we can be more aggressive in raising rates (i.e., using FER thresholds above 0%).

3.3.3.2 Channel Access Sublayer

The channel access sublayer (CAS) in Annex B to STANAG 5066 provides a mechanism for making and breaking physical links to other stations. The CAS also reports changes in status of the physical link to its client (the subnet interface sublayer [SIS]), and provides a conduit between the SIS and the DTS for sending and receiving datagrams.

More sophisticated channel access protocols, such as ALE and peer-to-peer channel sharing, are beyond the scope of Annex B.

3.3.3.3 Subnet Interface Sublayer

The subnet interface sublayer (SIS) in Annex A to STANAG 5066 provides the external interface to clients of the HF subnetwork service. Such clients bind to subnet access points (SAPs) on the SIS, which are analogous to ports in TCP (see Figure 3.13).

Clients of the HF subnetwork service interact with the SIS by exchanging interface primitives. These primitives carry client requests to the SIS, and responses and indications from the SIS to the client. Examples include requests to bind and unbind from an SAP, requests to establish and terminate links, requests to send datagrams, and indications of incoming datagrams.

Table 3.5 Data Rate Change Algorithm

Data Rate	*FER to Reduce*	*FER to Raise*
75	—	20%
150	50%	20%
300	50%	20%
600	50%	20%
1200	50%	20%
3200	50%	10%
4800	35%	5%
6400	20%	5%
8000	15%	2%
9600	5%	—

Session Types

The STANAG 5066 SIS offers three types of data sessions:

- A hard link is explicitly established, managed, and terminated by the client. ARQ is available on hard links, but is not mandatory.
- A soft link is not visible to the client, but is established by the SIS as needed to deliver datagrams placed in its queue. ARQ is used for reliable, in-order delivery. Soft links are terminated by the SIS when there is no more data for the linked destination, or as necessary to provide balanced service to other destinations.
- A broadcast session is established by the SIS to deliver non-ARQ data to one or more destinations. It may be permanent (for broadcast-only station), or established as needed.

The CAS makes and breaks the links upon request by the SIS.

Addressing

Nodes in the STANAG 5066 subnetwork are addressed using variable-size binary addresses (although all node addresses in a network are the same size). DTS headers contain a 3-bis addrese size field and as address field. Ths address field contains both the source and destination node addresses. Its size (in bytes) is specified in ths addrese size field. An address size of 000 indicates a 0-byte address field (i.e., implicit addressing). The maximum address field size is 7 bytes, or 28 bits per address. Thus, 32-bit IPv4 address cannot be used, and an address mapping from IP to 5066 addresses must be provided if Internet traffic is to be carried on the HF subnetwork.

3.3.3.4 Client Protocols

Annex F of STANAG 5066 defines client interfaces to the HF subnetwork service. For some client types, application-layer protocold that have been "tuned" for efficient operation over the HF subnetwork are also defined. All implementations of STANAG 5066 must present a subnetwork interface to IP clients on SAP 9, and must provide a TCP socket-server interface that listens on TCP port 5066. Other (optional) clients include the following:

- STANAG 4406 Annex E for tactical military message handling.
- A suite of HF-oriented e-mail clients: HMTP (derived from SMTP) for pushing mail through the network; HFPOP (derived from POP3) for pulling messages from servers; and CFTP, which is similar to HMTP but compresses messages for transmission;
- An operator orderwire (HF CHAT) client;
- Generic data transfer clients providing a character-oriented serial stream (COSS) service; a reliable connection-oriented protocol (RCOP) that uses STANAG 5066 ARQ; an unreliable datagram-oriented protocol (UDOP) that uses non-ARQ service; and an Ethernet interface that transfers the Ethertype and payload fields of Ethernet frames through the HF subnetwork.

The combination of CFTP, STANAG 5066 ARQ, and the advanced modems in this chapter enabled effective and interoperable use of email over HF radio links. In many user communities, email was the "killer app" that drove the rapid growth we have seen in HF data communications.

3.4 Channel Sharing

Our emphasis so far has been on point-to-point HF data links, but some applications share a single HF channel for peer-to-peer data distribution. In effect, we have a "local area" network of users sharing a single broadcast channel over distances of hundreds or thousands of miles. In this section we discuss the challenges of efficiently sharing such a channel.

3.4.1 Media Access Control Options

Approaches to sharing a communications channel are as old as human conversation, and we can draw inspiration for HF channel sharing mechanisms from human analogs. For instance, an informal discussion among friends over lunch often employs the following protocol:

- When I have something to say, I wait until the "channel" is free, and then start to speak.
- If two or more of us start speaking at the same time, we detect the collision, stop talking for a short, randomized time, and then, if another party has not seized the channel in the interim, try again.

This mechanism, similar to that employed in Ethernet, is called carrier sense multiple access within collision detection (CSMA-CD). However, it turns out to be difficult for colliding wireless stations to detect their collision, due to the disparity in transmitting and receiving power levels (as noted earlier). Therefore, wireless networks usually employ some form of collision *avoidance*, resulting in a CSMA-CA protocol. Collision avoidance usually takes the form of waiting a randomized number of "slots" after the channel becomes available before starting to transmit.

CSMA protocols are generally classified as "contention-based" media access control (MAC) protocols (see Figure 3.14). An alternative class of protocols avoids contending for the channel by instead scheduling use of the channel. Such

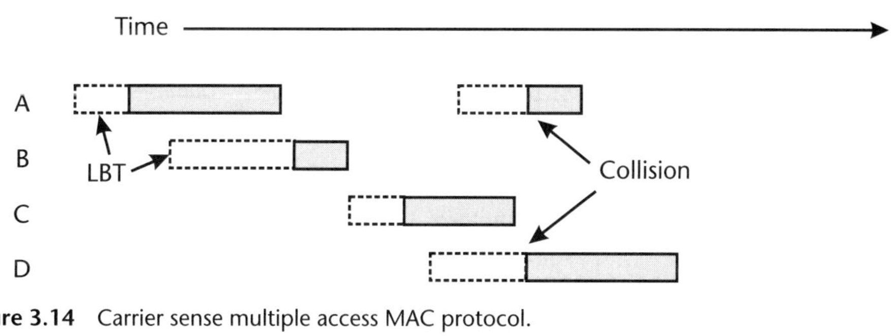

Figure 3.14 Carrier sense multiple access MAC protocol.

"contention-free" MAC protocols require some form of management of the channel. Two examples are time-division multiple access (TDMA) and token passing protocols:

- In the simplest form of TDMA, time on the channel is divided into non-overlapping time slots, and each member of the network is granted certain slots in which to transmit. No member transmits in any other member's slot, so there is no contention. However, when a member doesn't use its entire slot, the unused channel time is wasted (see Figure 3.15 station B). Other variants of TDMA include reservation mechanisms, which add some overhead but reduce wasted slot time.
- In a token-passing protocol, one network member at a time is granted the right to transmit; this right to transmit is conventionally termed the "token." When the network member holding the token has nothing more to send, or it has used the channel for the maximum permitted time, it passes the token to another member, ensuring that no channel time goes unused (Figure 3.16). Network members are organized into a ring for passing the token, but a broadcast channel carries data packets directly to their respective destinations; data does not need to be relayed around the ring.

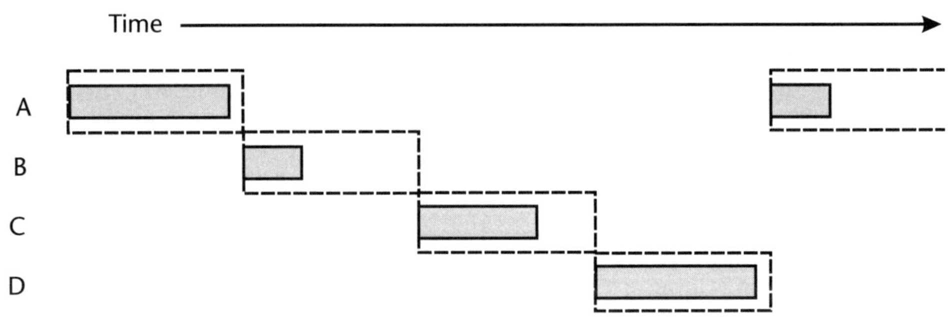

Figure 3.15 Time division multiple access MAC protocol.

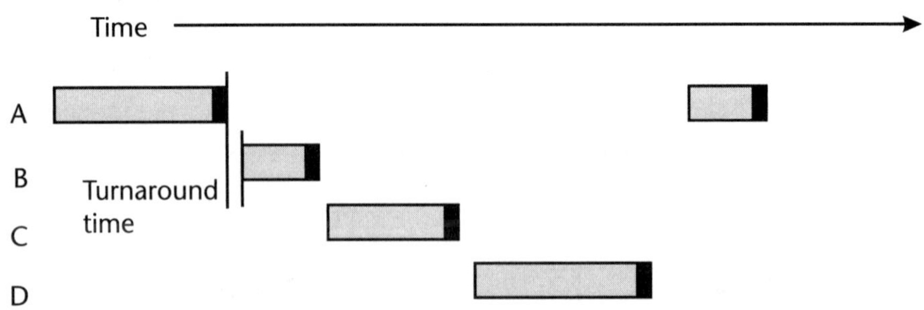

Figure 3.16 Token passing MAC protocol.

In a study presented at MILCOM 2003 [33], these alternative MAC protocols were evaluated for prospective use in an HF extended-line-of-sight LAN for naval battle groups. In this surface-wave application, the nominal network size was six nodes, spread over a diameter of 200 nautical miles. Traffic was a mix of operator-to-operator chat, email, and file transfers. The MAC protocol would be integrated with STANAG 5066 (modified as needed.) The physical layer was to be inventory HF radios, modems, and communication security (COMSEC) equipment.

Four candidate MAC protocols were developed and evaluated:

- A simple TDMA protocol with fixed slot sizes and allocations.
- A token passing protocol derived from the Berkeley Wireless Token Ring Protocol (WTRP) [34].
- The CSMA-CA protocol used in IEEE 802.11 [35], the distributed coordination function (DCF).
- A version of DCF modified for better performance in HF networks, dubbed DCHF.

The study evaluated the MAC protocols in terms of message latency in a lightly loaded network, and throughput achievable under heavy loading, in 5- and 50-node networks. Updated results are reported here for a 6-node network. In each case, data is sent at 6400 bps, with transmissions lasting up to 4.32 seconds (appropriate for MIL-STD-188-110B modems).

A key determinant of MAC performance was found to be the turnaround time, measured from the time the preamble of a packet arrives at the antenna of a node until the preamble of the response leaves the antenna of that node. In many wireless technologies, this time is measured in microseconds (or, at most, milliseconds), but for typical 2003-era HF radios, modems, and COMSEC, the measured turnaround time was about 2 seconds.

In networks using contention-free MAC protocols, only a single link turnaround time is required between transmissions from different nodes. However, DCF-based CSMA-CA protocols employ a request-to send (RTS) and clear-to send (CTS) handshake before exchanging a data packet and a link-layer Ack. These four packets require four link turnarounds for each data transmission. Therefore, as shown in the throughput graph in Figure 3.17, the CSMA protocols suffer greatly when turnaround times are seconds instead of milliseconds. As a result, the contention-free TDMA and token-passing protocols are clearly preferred when traffic is heavy.

Under light loading, we are interested in the delay to acquire the channel. Here, the CSMA protocols would be expected to excel, because they allow nodes to enter contention for the channel immediately. The contention-free protocols, by contrast, rotate channel access among the nodes; we therefore expect to wait, (on average), half of the respective cycle time to acquire the channel. The results for a six-node network are shown in Figure 3.18.

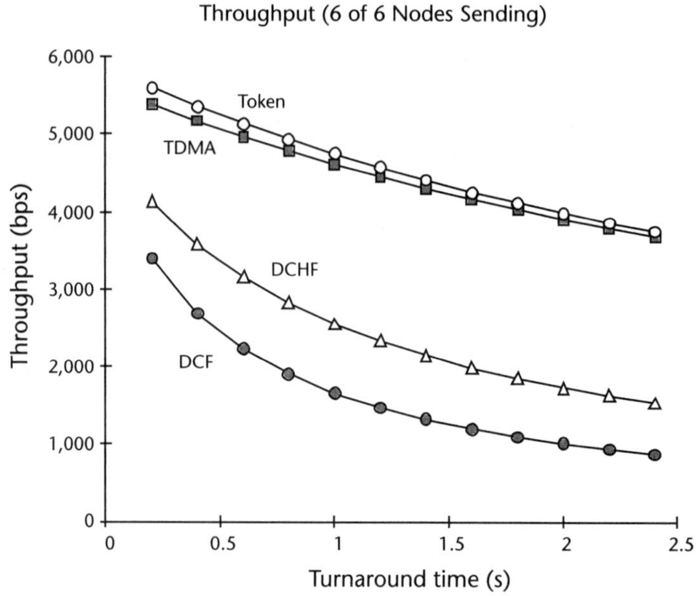

Figure 3.17 MAC protocol throughput comparison under heavy load.

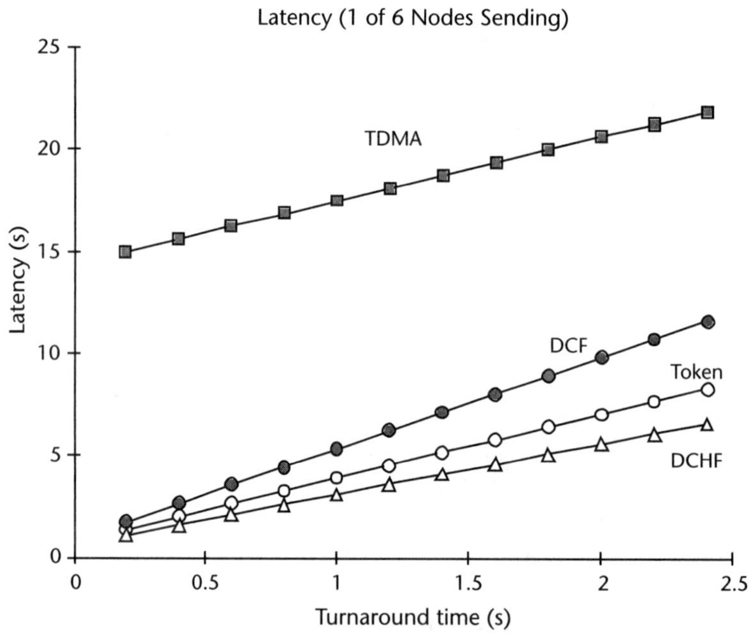

Figure 3.18 MAC protocol latency comparison under light load.

As expected, the DCF-based protocols have low latencies under light loads, especially when turnaround times are short[3]. TDMA latency is relatively high be-

3. DCHF is less sensitive to the turnaround time because DCHF omits the DIFS listen-before-transmit time before the contention-avoidance slots.

cause the cycle time does not vary with loading; under the light loading in this scenario, nearly all of the slots go unused. The token-passing cycle time under light loading is much shorter than that for TDMA, requiring only a turnaround time plus the time to send the token per node. In the six-node network, the access time for the token-passing MAC protocol was comparable to the CSMA results, but this would not hold for large networks.

The conclusion of this study was that a token-passing protocol offered good performance under both heavy and light loads. As a result, the HF Token Protocol derived from WTRP was fully developed, tested at sea by the U.S. Navy, and standardized in Annex L of STANAG 5066 (edition 3 is yet to be promulgated).

3.4.2 HF Token Protocol

The HF Token Protocol (HFTP) provides relatively efficient sharing of HF surfacwave channels, though the available throughput per station necessarily drops as stations are added to the network. When communications needs can be satisfied by multiple point-to-point links set up and released dynamically, higher throughput can be obtained through pooled use of multiple frequencies.

Key features of the token-passing protocol are presented here. Full details will be available in STANAG 5066 Edition 3, Annex L when this new edition is published.

- HFTP automatically forms a unidirectional ring of stations for circulating the right-to-transmit token (see Figure 3.19). Initially, when no ring is detected, stations enter "self ring" state in which they periodically solicit a successor.

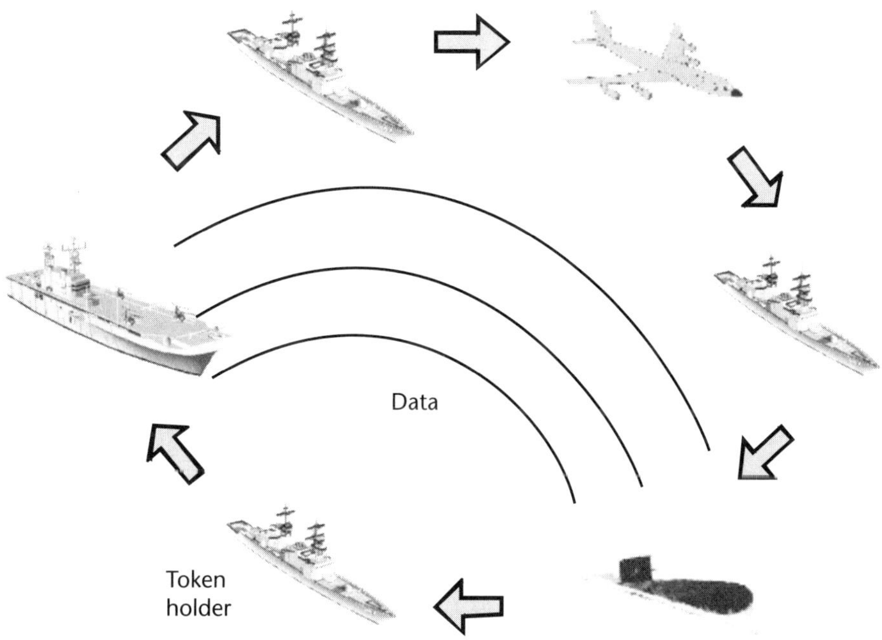

Figure 3.19 Token passing HF channel access protocol. (After [36].)

When another station responds to the solicitation, the two stations form a two-node ring.

- Stations that detect the presence of a ring, but which are not members of that ring, are not permitted to transmit. They must wait to be invited to join.
- Each member of a ring periodically solicits new stations to join the ring. If at least one new station responds, the soliciting station selects one responding station to insert into the ring as the soliciting station's new successor.
- A station holding the token is permitted to transmit only for a bounded time before it must pass the token to its successor in the ring. During the time it holds the token, a station may send data to any other stations in the network.
- Lost connectivity between adjacent stations in the token-passing ring is detected when the successor node is not heard to begin transmitting after the token is sent to it. Such lost connectivity is repaired either by re-threading the ring or by relaying the token around the lost link.
- Loss of the token is detected when the token has not circulated back to a station within a bounded time. A new token is created by the station detecting the loss.
- Duplicate tokens are detected and deleted.
- A station leaving the ring explicitly drops out by connecting its predecessor to its successor. If a station that has not departed cleanly becomes unreachable, this is detected and the station is removed from the ring by its neighbors.

This token-passing channel access protocol is becoming popular for naval battle group TCP/IP networks. The surface-wave channel simplifies finding a frequency that propagates to all network members. Experiments in using the token-passing protocol among stations linked by skywave links have been successful, but have also shown that finding a single working frequency can be challenging.

In Chapter 4, we address automatic frequency selection using automatic link establishment (ALE).

References

[1] Lin, S., and D. J. Costello, Jr., *Error Control Coding: Fundamentals and Applications, Second Edition*, Englewood Cliffs, NJ: Prentice Hall, 2004.

[2] Proakis, J. "*Digital Communications, Third Edition*", Boston: McGraw Hill, 1995.

[3] Wilson, S. "*Digital Modulation and Coding, First Edition*", Englewood Cliff, NJ: Prentice Hall, 1995.

[4] Chang, R., "Orthogonal Frequency Division Multiplexing," U.S. Patent 3488445, filed 1966, issued 1970.

[5] Malvar, H. "*Signal Processing with Lapped Transforms*, Norwood, MA: Artech House, 1992.

[6] Malvar, H., "Modulated QMF Filter Banks with Perfect Reconstruction," *Electronic Letters*, Vol. 26, 1990, pp :96--990.

[7] Linnartz, J., and S. Hara, "Special Issue on Multi-Carrier Modulation," *Wireless Personal Communications*, Kluwer, No.1--2, 1996.

[8] Vanderdorpe, L., "MMSE Equalizers for Multitone Systems without Guard Time," *Proceedings of EUSIPCO-96*, Trieste, Italy, September 10–13 1996, pp. 2049–2052.

[9] MIL-STD-188-110B, *Military Standard—Interoperability and Performance Standards for Data Modems*, U.S. Department of Defense, May 27, 2000.

[10] Kaiser, S., "Multi-Carrier CDMA Mobile Radio Systems—Analysis and Optimization of Detection, Decoding and Channel Estimation," Ph.D thesis, German Aerospace Center, VDI, January 1998.

[11] Perez-Alvarez, I., and I. Raos, et al., "Interactive Digital Voice over HF," *IEE Ninth International Conference on HF Radio Systems and Techniques*, University of Bath, UK, June 2003.

[12] Nieto, J. W., "Performance Comparison of Uncoded and Coded OFDM and OFDM-CDMA Waveforms on HF Multipath/Fading Channels," *SPIE Defense and Security Symposium*, Orlando, Florida, April 2005.

[13] Forney, G., "The Viterbi Algorithm," *Proceedings of the IEEE*, Vol. 61, No. 3, March 1973.

[14] Anderson, J., B. T. Aulin, and C. E. Sundberg, *Digital Phase Modulation*, New York: Plenum Press, 1986.

[15] Nieto, J., "Does Modem Performance Really Matter On HF Channels? An Investigation of Serial-Tone and Parallel-Tone Waveforms," *Nordic Shortwave Conference HF '01*, Fårö, Sweden: 2001.

[16] Nieto, J., "Constant Envelope Waveforms for Use on HF Multipath Fading Channels", *Proceedings of MILCOM 2008*, San Diego, CA: IEEE, 2008.

[17] Walker, W., and J. Sutherland, "Improved Signalling and Apparatus," U.S. Patent 4881245, filed 1987, issued Number 14, 1989.

[18] Brakemeier, A., "Criteria to Select Proper Modulation Schemes," *Proceedings of HF 95, Nordic Shortwave Radio Conference*, Fårö, Sweden, 1995.

[19] Wulich, D., "Reduction of Peak to Mean Ratio of Multicarrier Modulation Using Cyclic Coding," *Electronic Letters*, Vol. 32, No. 5, February 1996, pp 42-–433.

[20] Bauml, R., R. Fischer, and J. Huber, "Reducing the Peak-to-Average of Multi-carrier Modulation by Selected Mapping," *Electronics Letters*, Vol. 3, No. 22, 1996, pp 206-–2057.

[21] Cook, S., "Advances in High-Speed HF Radio Modem Design," *Proceedings of HF 95, Nordic Shortwave Radio Conference*, Fårö, Sweden, 1995.

[22] Giles, T., "A High-Speed Modem with Built-In Noise Tolerance", *Proceedings of the 6th International Conference on HF Radio Systems and Techniques*, York, UK, 1994.

[23] STANAG 4415, "Characteristics of a Robust, Non-Hopping, Serial-Tone Modulator/Demodulator for Severely Degraded HF Radio Links," *North Atlantic Treaty Organization, Edition 1*, December 24, 1997.

[24] STANAG 4285, "Characteristics of 1200/2400/3600 Bits per Second Single Tone Modulators/Demodulators for HF Radio Links," *North Atlantic Treaty Organization, Edition 1*, February 16, 1989.

[25] MIL-STD-188-110A, *Military Standard–Interoperability and Performance Standards for Data Modems*, U.S. Department of Defense, September 30, 1991.

[26] Widrow, B., and S. D. Stearns, *Adaptive Signal Processiion*, Englewood Cliffs, NJ: Prentice Hall, 1985.

[27] Johnson, R. W., M.B. Jorgenson and K.W. Moreland, "Error Correction Coding for Serial-Tone Transmission," *7th International Conference on Radio Systems and Techniques*, Nottingham, UK: 1997.

[28] Li, G., et al., "Coding for Frequency Hopped Spread Spectrum Communications," *Technical Report ECE-93-1*, April 1, 1993.

[29] Elliott, D., and K. Rao, *Fast Transforms Algorithms, Analyses, Applications*, London: Academic Press, 1982.

[30] Jorgenson, M. B., et al., "Polarization Diversity for HF Data Transmission," *7th International Conference on Radio Systems and Techniques*, Nottingham, UK: 1997.

[31] STANAG 5066, "Profile for High Frequency (HF) Radio Data Communications," *North Atlantic Treaty Organization*, 2008.

[32] Trinder, S. E., and A. F. R. Gillespie, "Optimisation of the STANAG 5066 ARQ Protocol to Support High Data Rate HF Communications," *Proceedings of MILCOM 2001*, IEEE, Tysons Corner, VA: 2001

[33] Johnson, E. E., M. Balakrishnan, and Z. Tang, "Impact of Turnaround Time on Wireless MAC Protocols," *Proceedings of MILCOM 2003*, IEEE, Boston: 2003.

[34] Ergen, M., et al., "Wireless Token Ring Protocol," *Proceedings of Systems, Cybernetics, and Informatics*, Orlando, FL: 2002.

[35] IEEE 802.11, "IEEE Standard for Information Technology—Telecommunications and Information Exchange Between Systems—Local and Metropolitan Area Networks–Specific requirements. Part 11: Wireless LAN Medium Access Control (MAC) and Physical Layer (PHY) Specifications," *IEEE*, 1999.

[36] Johnson, E., et al., "Robust Token Management for Unreliable Networks," *Proceedings of MILCOM 2003*, Boston: IEEE, 2003.

CHAPTER 4

Automatic Link Establishment

One of the key challenges in using HF skywave communications is finding a frequency that will support the desired voice or data traffic. We noted in Chapter 2 that the window of usable frequencies varies with the time of day, the seasons, the space weather environment, and the locations of the stations. As these dependencies became well understood, mathematical models of ionospheric propagation were developed and refined. Mainframe computer programs implementing these models could make propagation predictions, but these were not available for everyday use by most operators.

A simplified approach to frequency management was developed wherein frequency managers would identify frequency bands for use during specific hours, and then assign specific frequencies within those bands for use by operators. The simplest case was to specify a day frequency, a night frequency, and the times to switch between the two. This procedure certainly didn't provide 100% reliability, but it was easily understood and provided interoperability throughout a network. For critical HF radio links where the highest reliability was required, full-time, highly-skilled radio operators were employed to keep a link operational, changing frequencies as required by changes in ionospheric propagation.

Not surprisingly, after the introduction of satellites in the 1960s offered an alternative for beyond-line-of-sight communications, those who could afford the cost migrated away from the less-expensive but more difficult-to-use HF radios.

As microprocessors became widely available in the late 1970s, many formerly onerous tasks could be automated. This microprocessor revolution yielded personal computers powerful enough to run ionospheric prediction programs, such as IONCAP [1]. This was of some value to the remaining HF radio networks, but predictions could only offer statistical advice for frequencies to try; they could not reflect in real-time the space weather events that sporadically render such predictions irrelevant.

A more powerful use of microprocessors in HF radio was found when the microprocessor was no longer seen as a stand-alone device for *computing*, but was instead incorporated within the radio system to *control* the process of finding and using working frequencies. This process has since been named *automatic link establishment* (ALE) and is the subject of this chapter.

4.1 Introduction

In the late 1970s and early 1980s, engineers at leading HF radio companies realized that the job of finding a working frequency could be automated by embedding microprocessors in the radios they were designing. This made using HF radio easier. It was hoped that this would lead to wider use (and therefore more sales) of HF radios. Each manufacturer developed its own approach to ALE, and sales did indeed accelerate through the early 1980s.

Unfortunately, the proprietary systems from different manufacturers could not automatically establish links with each other. In fact, the speaker on an automated radio is normally muted until the radio is called using that specific manufacturer's protocol, so radios from different vendors would not even alert the operators that they were being called!

Proliferation of noninteroperable automated HF radio systems in U.S. federal agencies quickly became a concern to the National Communication System (NCS), an agency charged with ensuring that "functionally similar government telecommunications networks and facilities should be designed to provide the ability to rapidly and automatically interchange traffic in support of national security leadership requirements" [2]. In particular, the NCS was concerned that HF radios at sites across the United States could no longer be used to reconstitute government after a war or natural disaster destroyed other communications networks. (From first-hand reports to the authors, HF radios were kept in safes at emergency operations centers for such contingencies.)

In 1984, the MITRE Corporation undertook a study of U.S. government HF radio networks to assess the interoperability of existing and planned radio systems. As expected, the various proprietary systems were found to be noninteroperable, and MITRE was asked to suggest a way forward. In reviewing the capabilities of the existing ALE technologies, MITRE identified a succinct set of key functions that would contribute to a next-generation HF ALE standard for the U.S. federal government. These functions were organized into a step-by-step program of increasing capability [3] which became known as the "Stairway to Heaven" (Figure 4.1). These steps are briefly described below. The key functions in standard ALE will be described in detail later in this chapter.

- *Selective calling and handshake*: Each station is assigned a digital address (call sign) and implements a standard protocol for establishing links using those addresses.
- *Scanning*: A pool of frequencies is assigned for use in a network, and idle receivers cycle through those frequencies, listening for calls.
- *Sounding*: Stations in a network periodically transmit their call signs on the scanned frequencies, enabling the other stations to identify which frequencies are working.
- *Polling*: Stations conduct bidirectional tests of pool frequencies.
- *Connectivity exchange*: Stations implement a standard protocol for the exchange of connectivity information (for use in relaying or in routing protocols).

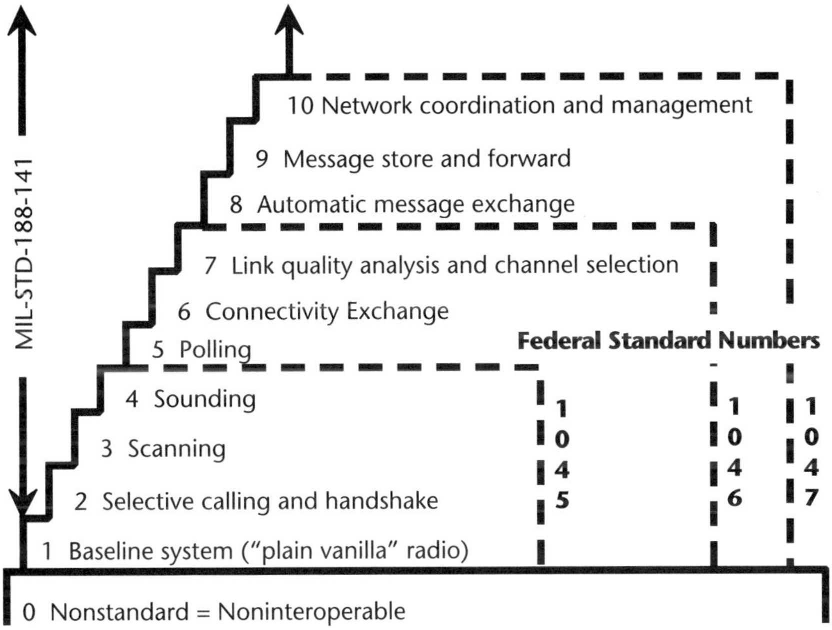

Figure 4.1 The "Stairway to Heaven."

- *Link quality analysis and channel selection*: Stations measure link quality (e.g., SNR) rather than a simple go/no-go for pool frequencies, store a database of these metrics, and use that database in selecting channels for placing calls.
- *Automatic message exchange*: Stations employ a standard protocol for the direct exchange of operator or user messages.
- *Message store and forward*: Stations are able to route messages indirectly to work around link outages.
- *Network coordination and management*: Stations implement a standard network management protocol.

In the course of this ALE study and functional analysis, the MITRE engineer, Gene Harrison, worked with many of the leading engineers in the U.S. HF industry. When the NCS tasked MITRE to develop a new federal standard for HF ALE, Harrison was able to merge some of the best thinking of these engineers (including ideas for their next-generation equipment) into the new standard, FED-STD-1045 [4]. By coincidence, the U.S. Department of Defense (DoD) was revising their HF radio standard at that time, and adopted the same ALE technology as a new Appendix A to MIL-STD-188-141A [5].

In terms of the (then popular) ISO seven-layer reference model, ALE was considered a data link layer function (Figure 4.2).

As Figure 4.2 illustrates, the ALE protocol employs protocol data units (PDUs) called *ALE words* for all of its link management and data transfer functions. An optional protection sublayer encrypts ALE words for protection from spoofing or other manipulation. Forward error correction (FEC) is then applied to these ALE

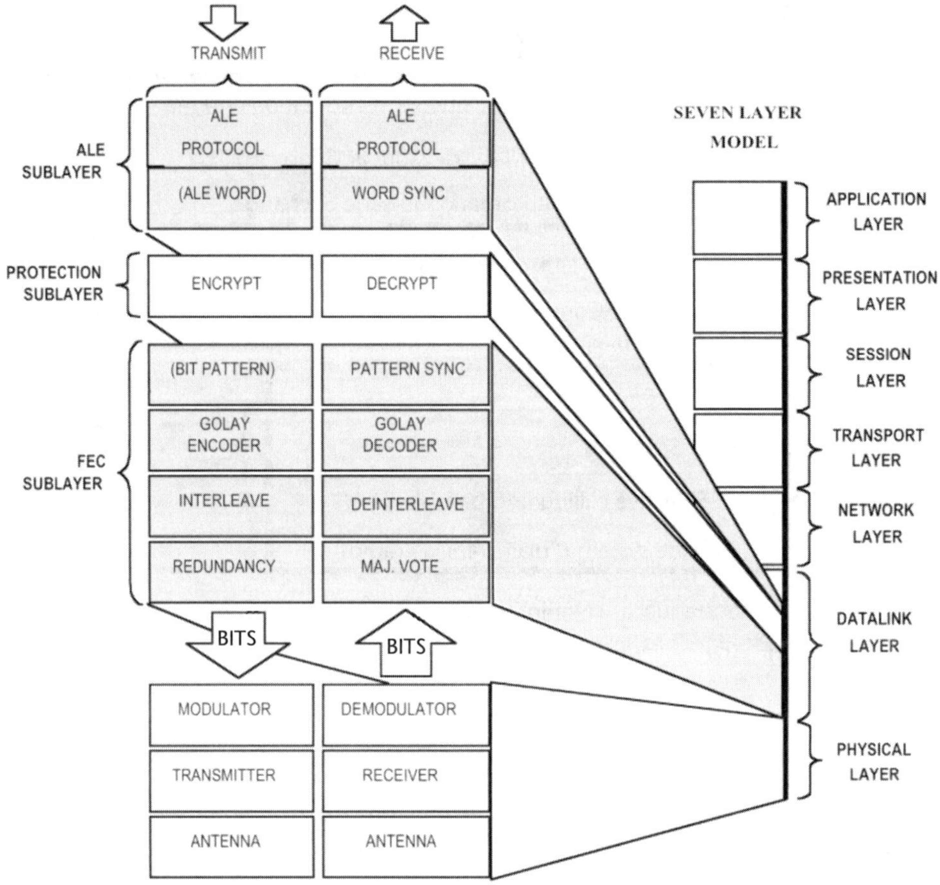

Figure 4.2 ALE sublayers. (From MIL-STD-188-141.)

words (encrypted or not), providing some protection from channel errors. The remaining sections of this chapter review the ALE signal structure and protocols, and conclude with a discussion of the linking protection scheme.

4.2 ALE Signal Structure

An ALE transmission is a contiguous sequence of ALE words called an *ALE frame*. The structure of the ALE frame will be described in Section 4.5. Here, we present the structure of the ALE word itself, the FEC applied to that word, and the modem used to carry ALE words over the HF channel.

4.2.1 ALE Modem

In Chapter 3, we discussed modem waveforms designed for carrying streams of data over the HF channels. A modem for ALE must cope with the same challenging channel characteristics, but must also efficiently convey short bursts of information, the 24-bit ALE words. Consequently, long interleavers and convolutional coding were ruled out for the ALE modem. Furthermore, the limited DSP technology available in the early 1980s prompted selection of a simple modulation that could be implemented using analog filters: 8-ary frequency shift-keying (FSK). One of eight orthogonal tones (Table 4.1) is sent in each symbol time, at a rate of 125 symbols per second. Each tone (symbol) represents three bits of data, so the raw data rate of this waveform is 375 bps.

The relatively long symbol period of 8 ms per tone keeps the data rate low, but permits this modem to cope with delay spreading without a need for the advanced signal processing found in adaptive equalizers (Chapter 3). Instead, we can simply discard the edges of the received symbols (where the intersymbol interference is found) and use only the middle of each symbol for demodulation. (The long symbol period also gives the ALE modem a characteristic warbling sound, which has become familiar to shortwave listeners who tune to ALE channels.)

4.2.2 ALE Word

The 24-bit ALE word is divided into two parts: a 3-bit preamble and a 21-bit data field (seen in Figure 4.3). In many cases, the data field carries three 7-bit ASCII characters representing all or a portion of a station address (call sign). The ALE word can also be used to convey orderwire commands over the HF link. In this case, the data section is partitioned in various ways, often using the first seven bits to specify a command and the remaining bits to carry parameters specific to that command.

Each ALE word is of one of eight types, as specified by the 3-bit preamble in that word. The word types and corresponding encodings of the preamble field are listed in Table 4.2.

The uses of the various types of ALE words will be presented in the discussion of the ALE protocols in Section 4.5.

4.2.3 Forward Error Correction

The goal of the FEC sublayer in Figure 4.2 is the error-free delivery of ALE words through the HF channel despite the noise, fading, and multipath spreading common

Table 4.1 8-ary FSK Modem for ALE

Tone (Hz)	*Bits*
750	000
1000	001
1250	011
1500	010
1750	110
2000	111
2250	101
2500	100

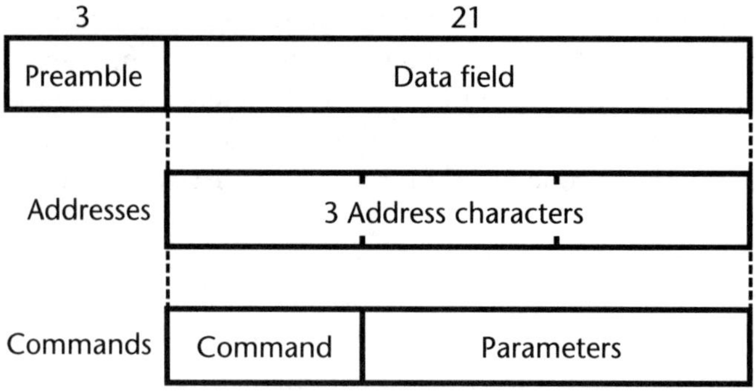

Figure 4.3 ALE word.

Table 4.2 ALE Word Types

Word Type	*Preamble*	*Use*
TO	010	Direct destination address
THRU	001	Group calls
TIS	101	Immediate source
TWAS	011	Immediate source; terminates link
FROM	100	Quick ID
CMD	110	Orderwire functions
DATA	000	Extension of preceding word(s)
REP	111	Duplicates preceding preamble

in skywave propagation. Four steps of FEC and related processing are applied to the ALE word to improve its chances of error-free reception. This ALE FEC processing expands the ALE word from 24 bits to a 147-bit redundant word. When sent using the ALE modem, the resulting 49 symbols occupy 392 ms on the air. This fundamental time quantum in the ALE system is usually abbreviated T_{rw} (duration of a redundant word).

4.2.3.1 Golay Coding

At the sending station, the first step in encoding the ALE word is to split it into two 12-bit halves. Each half is then encoded using the extended Golay (24, 12) code, a perfect block code that is capable of correcting up to 6 errors per 24–bit ALE word (3 bits in each Golay word). The FEC code generator polynomial is:

$$g(x) = x^{11} + x^9 + x^7 + x^6 + x^5 + x + 1$$

The generator matrix G derived from $g(x)$ contains an identity matrix I_{12} and a parity matrix P. That is, the Golay encoding is systematic, with the 12 bits from the ALE word sent unmodified, followed by 12 parity check bits. A simple approach for decoding this code, while adjusting its balance of error detection and correction, may be found in [6]. To assist the receiver in identifying word boundaries, the parity bits of the second Golay word are inverted before transmission.

4.2.3.2 Interleaving

As noted above, the span of error-correction efforts in the ALE waveform is limited to each individual ALE word. Thus, interleaving, which is used to spread channel errors for more effective processing by the Golay decoder, is applied only within the word (a 392-ms interleaver depth). In the original design, bits from the two Golay words were interleaved in a pseudo-random manner; however, this was found to offer little extra performance over a simpler perfect shuffle interleaving, so the latter was chosen for the standard.

4.2.3.3 Triple Redundancy

After interleaving the two Golay words, we have 48 bits of coded data. For added robustness, this coded/interleaved ALE word is sent three times in succession. At the receiver, majority voting is employed, both to correct some errors and to estimate the channel error rate. (Any nonunanimous vote indicates that at least one error has occurred.)

4.2.3.4 Stuff Bit

The original MITRE design for the FEC sublayer employed the three steps described so far. A fourth step was added to improve robustness for military uses. If a heterodyne tone (or a tone jammer) interfered with one of the ALE modem tones, this interference could be excised in the frequency domain, but we would still lose any bits sent using the affected tone. Triple redundancy wouldn't help because the 48-bit word divides evenly over the 3 bits per symbol, so every repetition of the word would produce the same 16-tone sequence. Any tone lost in the first instance would also be lost in the second and third instances.

A simple solution was to add a 49th stuff bit (always set to 0) at the end of the 48-bit coded/interleaved word. This causes the coded ALE bits to rotate through symbol boundaries, resulting in some time- and in-band frequency-diversity [7]. In practice, adding the 49th bit was found to improve SNR robustness in fading channels by roughly 1 dB.

4.2.3.5 Receive Processing

At the receiver, the first step in synchronizing with a new incoming signal is for the modem to identify 8-ms symbol boundaries. Once the modem declares the presence of ALE tones and timing, the next step is to achieve word synchronization. Identifying the boundaries between the redundant words is a cooperative process between the FEC sublayer and the ALE protocol sublayer. As each symbol (tribit) is delivered from the modem to the FEC sublayer, error correction and word sync processing proceed together as follows (see Figure 4.4):

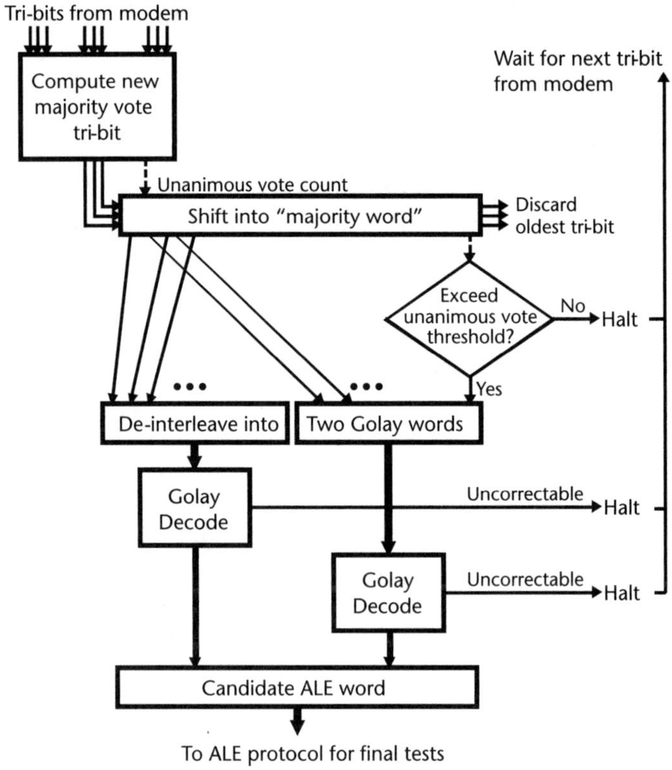

Figure 4.4 Word sync processing.

- Majority voting among tribits received at times T (the current tribit), $T - 49$, and $T - 98$ yields a majority tribit and a count of unanimous votes (0, 1, 2, or 3).
- This majority tribit is concatenated with the previous 15 majority tribits to form a 48-bit majority word (the 49th bit is discarded here). The total count of unanimous votes over the 16 majority tribits is compared to a threshold. If the unanimous vote total falls short of the threshold, it is unlikely that the 48-bit majority word is correctly framed, and processing halts until the next incoming symbol (tribit).
- If the unanimous vote threshold is met, the 48-bit majority word is de-interleaved into two 24-bit Golay words.
- The Golay words are decoded individually. If both words are correctable, the 12-bit results are concatenated to form a candidate 24-bit ALE word, and delivered to the ALE protocol for final determination of word sync. However, if either Golay word is uncorrectable, word sync will not be achieved at this symbol, and processing halts.

Once word sync is achieved, the FEC sublayer tasks of majority voting, deinterleaving, and Golay decoding are executed only after 49 new symbols have been received (not after every symbol).

4.3 ALE Addressing

The ALE system offers a very flexible addressing scheme, allowing up to 15 alphanumeric characters in an address. Characters in an address must come from the ASCII-38 alphabet, highlighted in Figure 4.5. This small subset of the ASCII table includes the digits, the uppercase letters, and two other symbols: ? and @. The former (?) is used as a wildcard character, and the latter (@) to fill unused positions in an ALE word when an address is not an even multiple of three characters (the capacity of an ALE word). The @ and ? characters are also used to form special-purpose addresses, as discussed in Section 4.5.5.

Every ALE station is assigned one or more ALE self addresses that apply to that station alone. These are termed individual addresses.

A single address can instead be assigned to a preprogrammed collection of stations. This is called a *net address*. Only those stations that are programmed to respond to a net address will know that it serves as a collective address; the format of net addresses is identical to that of individual addresses (see Table 4.3).

When it is desired to establish a one-to-many link to a collection of stations that have not been preprogrammed as a net, the group calling facility may be used. Here, the individual addresses of the desired station are listed in the call where there would normally be only an individual or net address.

Several special-purpose addressing modes are also defined for ALE systems.

- An *Allcall* address is used to call all available stations; stations that accept the Allcall do not respond, but stop and listen for traffic from the calling station.

BITS b_7 b_6 b_5 →					0 0 0	0 0 1	0 1 0	0 1 1	1 0 0	1 0 1	1 1 0	1 1 1
b_4	b_3	b_2	b_1	COLUMN → / ROW ↓	0	1	2	3	4	5	6	7
0	0	0	0	0	NUL	DLE	SP	0	@	P		p
0	0	0	1	1	SOH	DC1	!	1	A	Q	a	q
0	0	1	0	2	STX	DC2	"	2	B	R	b	r
0	0	1	1	3	ETX	DC3	#	3	C	S	c	s
0	1	0	0	4	EOT	DC4	$	4	D	T	d	t
0	1	0	1	5	ENQ	NAK	%	5	E	U	e	u
0	1	1	0	6	ACK	SYN	&	6	F	V	f	v
0	1	1	1	7	BEL	ETB	¢	7	G	W	g	w
1	0	0	0	8	BS	CAN	(	8	H	X	h	x
1	0	0	1	9	HT	EM	)	9	I	Y	i	v
1	0	1	0	10	LF	SUB	*	:	J	Z	i	z
1	0	1	1	11	VT	ESC	+	;	K	[	k	{
1	1	0	0	12	FF	FS	,	<	L	\	l	|
1	1	0	1	13	CR	GS	-	=	M	]	m	}
1	1	1	0	14	SO	RS	.	>	N	^	n	~
1	1	1	1	15	SI	US	/	?	O	?	o	DEL

Figure 4.5 The ASCII-38 character set for ALE addresses.

Table 4.3 ALE Address Types

Address Type	*Description*	*Example*
Individual	A single station	JOE
Net	Prearranged collection of stations	USA
Group	Ad hoc collection of individual station addresses	JOE BOB SAM
Allcall	Broadcast: no responses	@?@
Anycall	Broadcast: randomized responses	@@?
Wildcard	Matches multiple addresses: randomized responses	JO?

- An *Anycall* address also calls all available stations, but stations that accept the Anycall *are* expected to respond. To avoid collisions in their responses, the responding stations each select at random one of 16 slots following the call.
- The *Wildcard* address contains a '?' character that matches any alphanumeric character in the corresponding position in the address. As in the Anycall, stations called by a wildcard address select a random slot to respond to the call.

More details—including variations on these special-purpose addresses—were defined in Appendix A of MIL-STD-188-141 [5].

Most of the types of ALE words listed in Table 4.2 carry addresses. TO and THRU words specify the destination of a call, while TIS and TWAS specify the address of the calling station. Each ALE word carries up to three address characters. When using addresses longer than three characters, those extra characters are placed in DATA word(s) following the first (TO, THRU, TIS, or TWAS) word.

An important rule that governs the allowed sequences of word types comes into play when an address is longer than six characters: *successive ALE words cannot have the same preamble unless they are identical in content and function.* The REP (repeat) preamble is used when we would logically use the same preamble in adjacent words, but cannot due to this rule. Thus, to call a station using a 15-character address, the correct sequence of preambles is TO, DATA, REP, DATA, REP.

ALE stations are required to be able to recognize and respond to at least 20 individual self addresses, in addition to net addresses and special-purpose addresses.

4.4 Automatic Channel Selection

We turn now to the raison d'être of ALE: finding a suitable working frequency among two or more stations, and then establishing a link on that frequency. The first step, automatic channel selection (ACS) is discussed here, followed by link establishment protocols in Section 4.5.

ACS employs complementary mechanisms at the calling and called stations to automate what was once a tedious manual task. Of course, ALE can only succeed if the pool of assigned frequencies includes at least one that will work under each of the possible (not merely probable) conditions of time, season, and space weather. Therefore, the approach originally recommended for an ALE frequency pool was to span the entire HF spectrum, including some frequencies that shouldn't work under normal circumstances. Then, anomalous conditions (e.g., solar flares) would not result in extended outages. (As ALE networks have migrated from their backup role to being in constant use, spectrum planning has increasingly focused on providing sufficient channels in the bands likely to propagate.)

4.4.1 Scanning

The receiver's job is to be available for linking on all of the channels in the pool, although it is not continuously available on every channel. This is accomplished by repeatedly scanning those channels and listening for the tones and timing of the ALE modem. The scanning rate is determined by the time required for the receiving ALE demodulator to determine whether ALE tones and timing are present on a channel. The early (proprietary) ALE systems of the 1980s typically dwelled 500 ms on each channel, but recent implementations of the ALE standard routinely operate with dwell times of 100 ms.

Scanning is asynchronous in two respects:

- There is no scanning schedule for a network. Stations start scanning when powered on and when released from a link, and those times are generally unknown to other stations.
- The scanning dwell time of a station may be well-known, but this is only the *minimum* dwell time. When the modem finds ALE tones and timing on a channel, the scan will pause for up to 2 T_{rw} (784 ms) while attempting to achieve word sync (Section 4.2.3.5).

As a result, a station attempting to communicate with a scanning receiver generally will not know when that receiver will dwell on any particular frequency. Therefore, to have a high probability of capturing a scanning receiver, a transmission must last for 2 C T_{rw}, where C is the number of channels scanned.

4.4.2 Sounding

The calling station has the decision-making role in the ALE system: it is responsible for choosing the channel on which to place the call. This decision can be informed by propagation prediction programs, but a more direct approach was chosen for the ALE system. Every station maintains a database of recent *measurements* of propagation from other stations, and uses this database to select calling channels. (Of course, propagation predictions can also enter into the decision.)

Any station in the network that is accepting calls should assist in keeping other stations' ACS databases current by periodically transmitting its call sign (ALE address) on every channel. These sounding transmissions must last long enough so that every scanning receiver will have a chance to receive and measure the link quality from the sounding station to that receiver. As noted above, since we are attempting to reach asynchronously scanning receivers, the sound should ideally last

$2\ C\ T_{rw}$. However, this is quite conservative since scanning stations rarely pause for the full word sync time on every scanned channel. Operational ALE networks usually use a shorter sounding duration, since sounds can congest the channels if not used judiciously.

The format of a scanning sound is a continuous stream of ALE words that contain the address of the sounding station. The word type is usually TWAS, since this indicates that the sounding station will immediately leave the channel when it has completed its sound. If, instead, the TIS preamble is used, the sounding station is obliged to listen for calls *on that channel* for a short time after sounding.

4.4.3 Link Quality Analysis

A receiver (scanning or parked on a channel) that receives a sound will attempt to acquire word sync, and if successful, will read the address of the sending station. In the original MITRE design for FED-STD-1045, the fact that the sending station was heard on that specific channel would then be noted in a connectivity table. This table could then be used in selecting a channel when a link to that station was desired (Step 4 in the "Stairway to Heaven," Figure 4.1).

However, in the course of evaluating the ALE scheme for MIL-STD-188-141A, it was decided that Step 7 (link quality analysis(LQA)) would also be included in the basic ALE system. This meant that the *quality* of the link would be measured and recorded, not simply a "go/no go" record.

Several measures of link quality can be made in the ALE modem:

- A *pseudo-BER* can be estimated by noting the average number of nonunanimous votes per ALE word that occurred while acquiring word sync and reading the sounding station's address.
- SNR can also be measured by the modem. For example, if we have energy detection for each of the eight tone bins, we can compute the SNR during each symbol time as the ratio of energy in the active bin to the average energy in the other seven bins.
- Measuring multipath has also been considered, but this is more complex and is rarely done.

Thus, the connectivity table was upgraded to become an LQA database. When a call is to be placed, we can extract from that database a list of channels ranked by recent LQA scores for the desired destination. This is the automatic channel selection (ACS) approach commonly used in ALE networks.

4.5 ALE Protocols

When an operator directs the ALE system to establish a link to one or more destinations, the first step is selecting a suitable channel (or an ordered list of channels) to try. Then the stations execute an over-the-air protocol that brings them from their idle (scanning) state to a linked state; *linked* means that the stations are all tuned to the same channel and prepared to exchange traffic.

4.5.1 Frame Structure

Every ALE transmission is structured as shown in Figure 4.6.

- The address(es) of the station(s) to receive the frame are sent first so that every network member will know whether to stop scanning and receive the frame or resume scanning after reading the destination address(es).
- The station sending the frame is identified at the end of the frame.
- Any commands or messages are inserted in the optional center section of the frame.

When a called station is scanning, the calling station must capture that scanning receiver by sending the called station address repeatedly. This *scanning call* portion of the frame is the first part of any frame sent to a receiver that may be scanning (as seen in Figure 4.7). During the scanning call, only the first three characters of any ALE address are sent.

The next portion of the frame (always present) contains the complete address(es) of the called station(s), sent twice. This is termed the *leading call*. Finally, the *conclusion* of the frame contains the complete address of the calling station.

The contents and sequencing of frames differ for each type of calling protocol, as discussed below.

4.5.2 Individual Calling Protocol

An individual call sets up a point-to-point link from one calling station to one called station. The protocol comprises a three-way handshake (Figure 4.8). The calling

Figure 4.6 ALE frame structure.

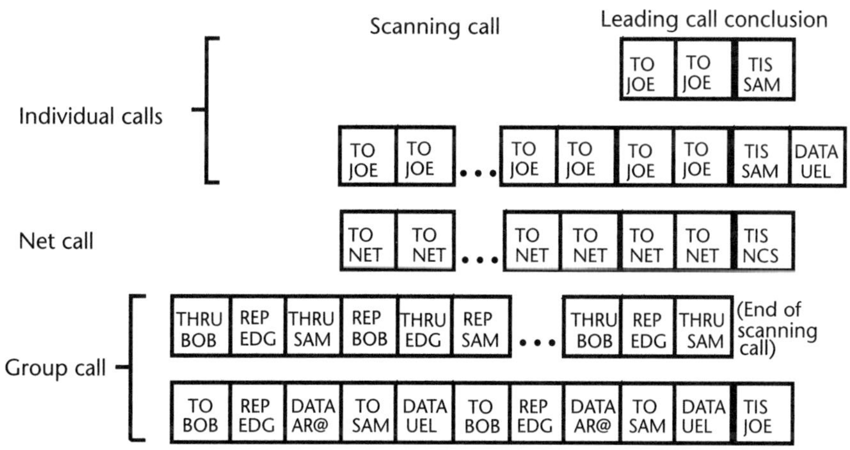

Figure 4.7 Example ALE frames.

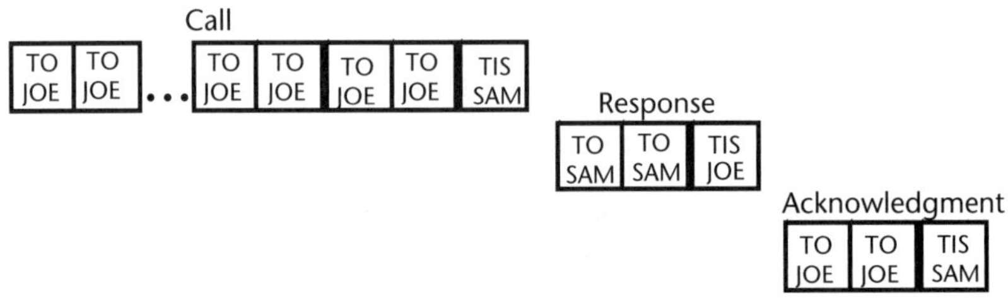

Figure 4.8 Individual calling protocol.

station sends a *call* frame, which begins with a scanning call if the called station may be scanning. Every ALE word in the scanning call is a TO word. Use of the TO preamble indicates that the word contains the first three characters of the called station's address.

The called station, if it is scanning, will arrive on the selected channel at some point during the scanning call, read its address in a TO word, and recognize that it is being called. Any other stations that read the TO word will find that the address does not match any of their self addresses and will depart to scan other channels.

The called station does not know which station is calling until the conclusion of the call. The conclusion begins with either a TIS or a TWAS word. The TIS preamble indicates that the calling station wishes to establish a link. A TWAS preamble identifies the calling station, so the link quality analysis of the call can be correctly entered in the receiving station LQA database, but indicates that the calling station will return to scanning after sending the frame.

The called station does not know the length of the calling station address until it has either received five ALE words in the conclusion, or the signal is lost after receiving at least one word of the conclusion[1]. Thus (except in the case of five-word addresses), the called station must wait one T_{rw} past the end of the transmission to be sure that the transmission is ended.

After successfully receiving a call, the called station will send a *response* frame, which is addressed to the calling station. Before sending the response, the called station may need to tune its antenna coupler. The calling station allows some extra time for this (a programmable parameter). If the calling station does not receive a timely response, it automatically aborts the linking attempt and either returns to scan or retries the call on another channel.

If the calling station receives a timely and correct response to its call, it now knows that the selected channel is propagating in both link directions. However, the called station does not know that the channel propagates back to the calling station. Therefore, the caller sends a third transmission, an *acknowledgment*, to the called station. This completes the link establishment protocol. At this point, the speakers are unmuted for voice traffic, or a data link protocol is engaged for passing data.

Once stations are linked, they both start a wait for activity timeout (nominally 30 seconds) that will return both stations to the idle (scanning) state if neither station transmits during the period of the timeout. This timeout is stopped whenever either station is transmitting, and restarted from 0 at the end of each transmission.

1. Signal loss is detected either by the radio or by the Golay decoder.

When a station returns to scan, it may announce this by sending a TWAS-concluded frame.

4.5.3 Net Calling Protocol

A net call establishes a multipoint link among multiple stations. Recall that a net address is programmed to be recognized by many stations. When a net address is called, using the same call frame structure as in an individual call, all of those stations will prepare to respond. How do we avoid collisions among those responses? By assigning those responses to individual time slots following the call.

When an address is programmed as a net address, it will be accompanied by a slot wait timer value (different for each net member) that determines how long the called station will wait *after the end of the call frame* before sending its response. A simple case of such slotted responses is shown in Figure 4.9.

The first slot after the end of the call, Slot 0, is always reserved (e.g., for tuning antenna couplers) and is not used for responses. After Slot 0, each net member responds in its assigned slot. After all of the slot times have passed, the calling station will send a collective acknowledgment (if any stations have responded). This completes the three-way handshake.

4.5.4 Group Calling Protocol

When a link is desired to a group of stations that is not preprogrammed, a group call may be used. This presents some interesting problems:

- A single address has not been programmed to refer to this group of stations, so the call must list their individual addresses. However, if a scanning station detects a call to an address that is not one of its self addresses, it will ignore that call and return to scanning without waiting to see if its address comes later in the list.
- We will need slotted responses, but slot wait times have not been assigned.

The group calling protocol addresses these concerns as follows:

- The scanning call portion of the call frame carries the first three characters of each called station address in THRU words (not TO words). However, we

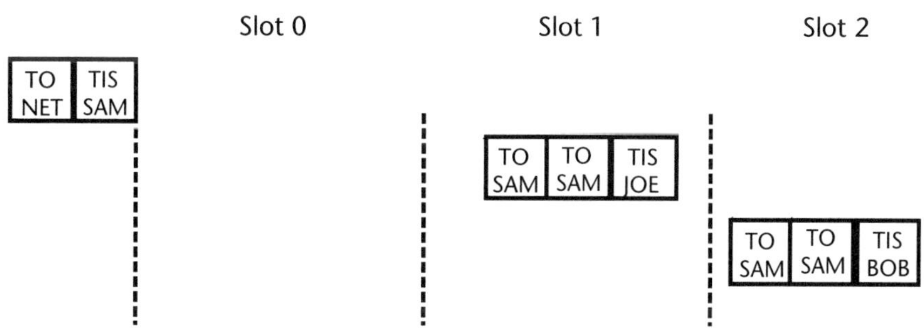

Figure 4.9 Slotted responses.

cannot repeat preambles when the ALE words are not logically identical, so THRU preambles must alternate with REP preambles (as shown in Figure 4.7). A scanning station that decodes THRU or REP words when it synchronizes to an incoming ALE transmission is obliged to continue to decode ALE words until it encounters the first three characters of a self address or it has seen a complete cycle of addresses.

- The leading call portion of a group call uses TO and REP words, and includes the *full* addresses of all called stations (sent twice, as seen in Figure 4.7).
- Response slots are computed on the fly: the last station named in the leading call will respond in Slot 1, the preceding station in Slot 2, and so on in *reverse* order of the list in the leading call. (As in the net call, Slot 0 is not used for responses.)

As in the net call, the calling station will send an acknowledgment after the last response slot.

4.5.5 Other One-to-Many Calling Protocols

Three other one-to-many calling protocols are defined for the ALE system:

- An *Allcall* uses an address of the form @?@. All stations that receive an Allcall stop and listen, but do not respond. If some character other than "?" is placed between the two "@" characters, the call is a *selective Allcall.* Only stations whose address *ends* in the character between the two "@" characters are called, and should stop and listen. Other stations ignore the selective Allcall.
- An *Anycall* uses an address of the form @@?. A *selective Anycall* uses an address with some character other than "?" after the two "@" characters, and only stations whose address ends in that character are being called. Others ignore the selective Anycall.
- A *Wildcard* call is any call other than an Allcall or an Anycall that contains a "?" in the address.

Anycalls and Wildcard calls use slotted responses, but responding stations cannot compute slot assignments for these types of calls. Instead, sixteen slots are always present, and responding stations choose one of those slots at random for sending the response. An unused Slot 0 immediately follows the end of the call, and the response slots (numbered 1 through 16) follow Slot 0. As usual, an acknowledgment follows the last response slot.

4.5.6 Timing

The timing characteristics of the radio, antenna coupler, and so on affect the operation of the protocols. Programmable parameters are provided for many of these times (detailed in MIL-STD-188-141, Appendix A [5]). It is important in programming an ALE network to set these parameters identically in all network member ALE systems.

4.5.7 ALE Performance Requirements

The purpose in defining the ALE standard was to ensure interoperability among government HF radio systems, but users also desired minimum *performance* requirements. These fall into two categories: occupancy detection and linking probability.

4.5.7.1 Occupancy Detection Requirements

A potential problem in any automated communication system arises when it may interfere with other users of the medium. The early implementations of the ALE standard all listened before transmitting on the calling channel, but some only detected (and deferred to) ALE signaling. As a result, these systems earned the scorn of users whose voice conversations were interrupted by the infamous warbling tones of the ALE modem.

When a radio operator on Air Force One (the U.S. presidential aircraft) complained of such interference, the military standards committee quickly added requirements for ALE systems to reliably detect voice traffic, as well as ALE and data modem transmissions, and to defer transmitting on occupied channels. These requirements are listed in Table 4.4, and are tested using standard recordings of all three types of traffic.

All ALE systems are required to listen for a dwell time of two seconds before transmitting a call or a sound, and to detect occupancy with the probabilities listed in Table 4.4. False detection probability cannot exceed 1%.

4.5.7.2 Linking Probability Requirements

The ability of ALE systems to establish links is tested in three channel conditions, as listed in Table 4.5. The unusual specifications for multipath spread were selected to preclude nulling of any of the ALE tones (which would occur if the usual 0.5 and 2.0 ms settings were used in standard Watterson model channel simulators).

4.5.8 Orderwire Functions

Numerous control and messaging functions have been defined that use ALE CMD words sent in the optional message section of an ALE frame. Very few of these

Table 4.4 Occupancy Detection Requirements

Waveform	*SNR (dB in 3 kHz)*	*Dwell Time (s)*	*Detection Probability*
ALE	0	2.0	80%
	6	2.0	99%
SSB Voice	6	2.0	80%
	9	2.0	99%
Serial-tone PSK modem	0	2.0	80%
	6	2.0	99%

Table 4.5 Linking Requirements (SNR in 3 kHz)

Probability of Linking	*AWGN Channel*	*"Good" Channel*	*"Poor" Channel*
≥ 25%	−2.5 dB	+0.5 dB	+1.0 dB
≥ 50%	−1.5 dB	+2.5 dB	+3.0 dB
≥ 85%	−0.5 dB	+5.5 dB	+6.0 dB
≥ 95%	0.0 dB	+8.5 dB	+11.0 dB
Multipath (ms)	0.0	0.52	2.2
Doppler spread (Hz)	0.0	0.10	1.0

are mandatory, and so are not universally implemented. Two of the mandatory functions are illustrative of the capabilities:

- The LQA command, carried in a single CMD word, allows stations to report to each other the pseudo-BER, SNR, and multipath that they have measured on transmissions from each other. This may be useful when links are nonreciprocal (propagate better in one direction than the other), often due to local interference on some channels.
- Automatic message display (AMD) provides a low-overhead operator-to-operator text message capability using the ALE signaling. An ALE system that receives an AMD message is required to display it to the operator and to store it for later review. AMD messages are limited to 90 characters, using an alphabet of upper-case letters, digits, and the punctuation symbols.

The sending station may optionally insert its address before an especially long message section. This *quick ID* uses a FROM word (extended as usual with DATA and REP words as needed).

4.6 Linking Protection

In response to concerns about the security of ALE systems against spoofing, a technique called *linking protection* (LP) was developed to frustrate unauthorized attempts to interact with ALE systems, either by establishing unauthorized links or interfering with the establishment of legitimate links. Note that LP does not address jamming or similar techniques, which are best countered by TRANSEC, nor is it intended to replace the COMSEC function of traffic protection. LP protects the *linking* function, including related addressing and control information.

4.6.1 Requirements

The approach chosen for LP is to authenticate ALE transmissions before they are accepted for action. A cryptographic technique was desired because it would provide strong authentication. The following requirements were agreed to guide the design of the LP technique:

- *Transparent to ALE protocols.* The first requirement was that the linking protection mechanism be completely transparent to the ALE protocols, so that it could be added modularly to any system that implements ALE. This means that the tones, timing, redundancy, interleaving, FEC, and protocols must be identical for the protected and unprotected modes of operation. In particular, linking protection could not require the transmission of any additional bits for synchronization or similar purposes.
- *Self-synchronizing.* Because a principal need for linking protection is in denying an adversary the ability to establish unauthorized links, the linking protection mechanism must be effective when radios are scanning; this is when links are normally established. The mechanism must therefore be self-synchronizing so that radios arriving on-channel after the start of a transmission can acquire crypto sync and begin checking for transmissions addressed to them.
- *Minimum impact on scanning dwell time.* Unauthorized transmissions should ideally cause a scanning receiver to pause no longer than normal on a channel carrying deceptive signaling. Thus, a scanning receiver must be able to gauge the authenticity of received transmissions in the time usually required for word sync.
- *24-bit block operation.* The basic unit of ALE transmissions is a 24-bit ALE word. The linking protection mechanism therefore needed to map 24-bit words into 24-bit words that can be transmitted immediately. Likewise, when a 24-bit word is received, the LP mechanism needs to be able to decrypt that word immediately, without a need to receive more bits. This is necessary for word sync acquisition.
- *Channel- and time-varying.* The ciphertext produced from identical plaintext must vary from channel to channel at any time instant, and must also vary periodically on the same channel, so that protected stations are minimally affected by tape recorder attacks.
- *Moderate computational requirements.* The computational complexity of the LP scheme needed to be implementable within the power and timing constraints of 1990 field radios.
- *Unclassified algorithm.* An unclassified cryptographic algorithm was desired for at least some applications of LP, so that a protected radio would not require the physical security needed for high-grade COMSEC devices.

4.6.2 LP Technique

The technique chosen for HF LP was time- and frequency-dependent *encryption* of ALE words using a 24-bit block algorithm. This provides authentication at the receiver because only a network member can produce an encrypted ALE word that will decrypt correctly.

The time of day (TOD) and operating frequency are incorporated into the encryption process through a seed that is used by LP algorithms in similar fashion to the cryptographic key. The standard seed format (see Figure 4.10) contains the following fields:

- *Date*: 4 bits for the month and 5 bits for the day of the month;
- *Protection Interval (PI)*: 11 bits for minutes since midnight, 6 bits for seconds in the current minute;
- *Word*: a count of ALE words encrypted during this PI (see below);
- *Frequency*: the nominal frequency carrying the protected transmission, in binary-coded decimal (BCD). The digits range from hundreds of MHz down to hundreds of Hz.

An important consideration in TOD-based cryptography is that the network must be synchronized to roughly the same time quantization as is used in encryption. For example, if stations in the network are synchronized to within one second of each other, we should use TOD quantized to one second for encryption. Then, when a station receives a protected transmission, it is known that the TOD at the sender is within one second of the TOD at the receiver. The receiver would therefore need to try decrypting the transmission using the following TOD values:

- Receiver's current TOD;
- Receiver's current TOD + 1 s (the transmitter could be ahead);
- Receiver's current TOD – 1 s (the transmitter could be behind).

If the TOD quantization used in encryption was instead 100 ms, the receiver would need to try 21 TOD values in 100 ms steps from its TOD – 1 s through its TOD + 1 s.

The time quantization used in LP is termed the *protection interval* (PI). The PI field in the seed contains the current time in minutes and seconds since midnight, quantized by the protection interval in use in the network. For example, if the protection interval is 2 seconds, then the PI seconds field will always be an even number.

Protection intervals are always at least one second in duration, so multiple ALE words will be encrypted in each PI. The security of the LP technique requires that a different seed be used for each ALE word; therefore, we have a word number field in the seed that is incremented for each succeeding word in a PI. The word field is reset to 0 at the start of the each PI.

Once a receiver is synchronized with the LP process in a transmission, the sequence of word numbers is easily followed. However, when a receiver first arrives

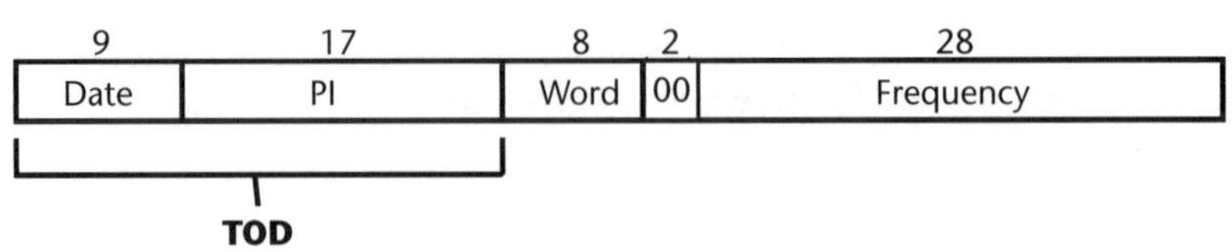

Figure 4.10 Linking protection seed.

on a channel carrying a protected scanning call, what word number should the receiver assume was used by the transmitter? To avoid the need to try a large range of word numbers, a special technique is used during the scanning call: the transmitter simply alternates between word 0 and word 1. If the receiver successfully decrypts a received word using word number 0, the next word must use word 1.

4.6.3 Application Levels and Algorithms

Recognizing that users may desire different combinations of cost, overhead, and security for their applications of LP, a range of standard application levels was defined. Every level is able to interoperate with less protected levels *when so directed by an operator*. Table 4.6 defines the PI duration (synchronization requirement) and the type of cryptographic algorithm used in each application level.

Note that application level 0 (commonly abbreviated AL-0) is unprotected ALE.

Special 24-bit cryptographic block algorithms were developed for LP. The unclassified LATTICE algorithm [8] was designed for efficient software implementation in 1990-era microcontrollers. It was eventually approved for export. The Type II and Type I algorithms are more tightly controlled, and are implemented in special hardware modules. The Type I module requires COMSEC-level physical security, and so is rarely used.

4.6.4 Time Synchronization

Synchronizing station time bases across a network is simple if every radio has access to GPS time. However, when HF radio is intended for backup beyond-line-of-sight communications when satellites have been disabled, it is necessary to provide organic means to synchronize an HF network, without recourse to GPS. Therefore, a suite of time exchange protocols was developed for HF networks that use LP. (Recall that unprotected networks are asynchronous and do not require synchronized time bases.)

4.6.4.1 Time Quality

The concept of a *time uncertainty window* is fundamental in time distribution for linking protection. It measures the amount of uncertainty in a time source; for

Table 4.6 Link Protection Application Levels

Application Level	*PI Duration*	*Algorithm*
0	(no LP)	(none)
1	60 s	LATTICE
2	2 s	LATTICE
3	2 s	Type II
4	≤1 s	Type I

example, how far that timebase may have drifted from coordinated universal time (UTC). The size of a time uncertainty window at a station is determined by the accuracy and precision with which that timebase was last set, plus a term that grows with time at a rate determined by the stability of that time base.

For example, if a station's LP clock is set with a timing uncertainty of ±10 ms, its time uncertainty window is then set to 20 ms (total time uncertainty). If its oscillator has a stability of ±10 ppm, this uncertainty window grows at a rate of 72 ms per hour.

Now, assume that this station sends time to another station 3 hours after its clock was last set. The time uncertainty window has grown to 236 ms, so the station receiving time will need to start its time uncertainty window at this size, plus any additional timing uncertainty that arises in the time transfer. Unless we know the propagation delay over the HF channel, we should add 70 ms of uncertainty for skywave propagation. If there is 100 ms of processing time uncertainty at the distant station, the total time uncertainty window at that distant station will start at 236 + 70 + 100 = 406 ms.

Instead of using a lot of bits to report the time uncertainty window at a time source, the time exchange CMD instead quantizes uncertainty into 8 levels of *time quality*. The upper bounds on time uncertainty for each time quality level are listed in Table 4.7.

Reworking our example, the time source would report that its time is quality 3, and the receiving station will start its time uncertainty window at 500 + 70 + 100 = 670 ms.

The time uncertainty window concept is useful in computing how often a station must resynchronize its timebase to stay synchronized within the PI duration of its network. Continuing with our example, the station that received time over the air starts its time uncertainty window at 670 ms. If its timebase stability is ±10 ppm, how long can it go before it needs to request an update to maintain AL-2 synchronization? The maximum time uncertainty for AL-2 is 2000 ms, so our station must request an update after (2000 – 670) / 72 = 18 hours.

4.6.4.2 Time Service Protocols

A range of time delivery protocols is specified in Appendix B of MIL-STD-188-141 [5], covering several situations:

Table 4.7 Time Quality

Time Quality Code	*Time Uncertainty Window*
0	none
1	20 ms
2	100 ms
3	500 ms
4	2 s
5	10 s
6	60 s
7	unbounded

- When the time server and time requester are both synchronized to within the PI in use in their network, a protected time exchange handshake can be used to deliver time securely, relying upon the cryptographic protection afforded by the LP algorithm.
- A station that is not synchronized cannot use a protected handshake, but can instead send an unprotected request for time. This request includes a random nonce to help authenticate the response. A time server responds to an unprotected request with the correct time, its time quality, and an authentication word. The authentication word is produced by encrypting the nonce using the network key and the reported time. If the requester validates this authentication word, the time response is probably authentic.
- Protected and unprotected (but authenticated) time broadcasts are also defined.

4.6.4.3 Time Iteration Protocol

It is possible to measure the propagation delay on the HF channel, and so remove this element of time uncertainty from the time delivery process. This is accomplished by exchanging delta time reports, which contain the differences between measured and reported timestamps. Using these, stations obtain samples of the offset between their respective local times plus residual randomness. The iteration continues until the resulting time uncertainty is reduced to an acceptable window [9]. This protocol has not been standardized.

References

[1] Teters, L. R., J. L. Lloyd, G. W. Haydon, and D. L. Lucas, "Estimating the Performance of Telecommunication Systems Using the Ionospheric Transmission Channel–Ionospheric Communications Analysis and Prediction Program User's Manual," *Report NTIA 83-127, National Telecommunication and Information Administration*, Boulder, CO, 1983.

[2] Reagan, R., National Security Decision Directive Number 97, "National Security Telecommunications Policy," The White House, Washington, DC, June 13, 1983.

[3] Harrison, G., "Functional Analysis of Link Establishment in Automated HF Systems," *Working Paper 86W00015*, MITRE Corporation, McLean, VA, December 1985.

[4] Federal Standard 1045, *Telecommunications: HF Radio Automatic Link Establishment*, General Services Administration, January 24, 1990

[5] MIL-STD-188-141A, *Interoperability and Performance Standards for Medium and High Frequency Radio*, September 15, 1988. (This version has been superseded by MIL-STD-188-141C, dated 25 July 2011.)

[6] Johnson, E. E., "An Efficient Golay Codec for MIL-STD-188-141A and FED-STD-1045," *Technical Report NMSU-ECE-91-001*, NMSU, February 1991.

[7] Johnson, E. E., "Addition of a 49th Bit to the MITRE HF ALE Waveform," *Technical Report PRC-EEJ-88-002*, NMSU, March 1988.

[8] Johnson, E. E., "A 24-Bit Encryption Algorithm for Linking Protection," Technical Report NMSU-ECE-89-027 (Restricted Distribution), 1989. (Also available as "USAISEC Technical Report ASQB-OSO-S-TR-92-04.")

[9] Johnson, E. E., "Time Iteration Protocol for TOD Clock Synchronization," **NMSU**, 1992.

CHAPTER 5

Third-Generation Technology

By the mid-1990s, the standard ALE system had become notably successful, initially in the United States and then globally. When the ALE standard was developed in the 1980s, few U.S. government agencies were using HF radio for anything besides backup and emergency communications. The standard, therefore, emphasized interoperability rather than efficiency in using the spectrum. However, ALE made HF radio much easier to use, so its use naturally grew rapidly. Concern began to grow over the increasing congestion of the very limited number of HF frequencies available worldwide. Some features of the ALE system exacerbated this congestion were:

- The ALE system's asynchronous mode of operation requires a lengthy scanning call to ensure that scanning stations will receive the call. The length of the scanning call increases with the number of frequencies scanned.
- Automatic channel selection requires that all stations that may be called must transmit sounds on every channel. Once again, the length of each sound is proportional to the number of channels scanned.
- As the amount of traffic in a network increases, more channels are needed to accommodate that traffic. Thus, we find that both channel congestion and overhead utilization of transmitters increase quadratically with network traffic. This effectively bounds the traffic level and number of stations that can be accommodated by the ALE system of the 1980s.

Other shortcomings of the ALE system were also noted:

- The time required to complete an ALE call was 10 to 20 seconds, which seemed excessive to some users (although it was comparable to the time to place an international direct-dial call through the telephone network).
- The ALE system was designed primarily to support voice service, but data transmission over HF radio had become increasingly important. Data was sent using PSK waveforms, so some system architects objected to using the ALE FSK modem for channel selection, fearing that channels suitable for a PSK waveform might be rejected by FSK-based measurements (perhaps because the data modems could operate with lower SNR than could the 8-FSK ALE modem).
- The DSP technology of the 1990s was far advanced over that available in the 1980s. Robust waveforms using Walsh coding could communicate at perhaps 10 dB lower SNR than the ALE system could establish links. While such

links would not be especially useful for voice service, they could certainly carry messages and small volumes of data.

Thus, it was clear that a new generation of ALE technology was needed. At the urging of the DoD, a joint effort among industry and academic engineers was initiated, seeking once again to bring together the best available ideas into a new standard. To eliminate confusion in the HF radio community as this new development commenced, the following terminology was adopted:

- The original ALE technologies developed independently by industry were termed *first generation ALE.*
- The standardized, interoperable system in MIL-STD-188-141A and FED-STD-1045 was now called *second generation* or *2G ALE.*
- The new project would produce a standard to be called *third generation* or *3G ALE.*

The goals for the 3G HF technology project were to produce a technology that established links faster and at lower SNR, used the spectrum more efficiently so it could support more stations and heavier traffic loads, used similar modem waveforms for link establishment and traffic, and that efficiently supported Internet applications.

Quantitatively, the 3G technology was to achieve order-of-magnitude improvements over the 2G technologies in three dimensions: reduce the SNR required to set up a link by 10 dB, accommodate 10 times as many stations in a network, and improve data traffic throughput in a network tenfold *in the same allocation of spectrum.*

In this chapter, we present the resulting suite of 3G HF technology, along with measurements and analysis of its performance. Full technical specifications of the 3G suite may be found in NATO STANAG 4538 [1].

5.1 Introduction to the 3G HF Technology Suite

The project to develop a new generation of HF radio technology offered a blank slate for the developers. As a result, the various capabilities that had evolved separately in an ad hoc manner in the previous generation could now be designed as parts of an integrated suite. Figure 5.1 depicts this integrated family of technology in the context of a protocol stack.

The 3G HF system provides an integrated communication service to applications ranging from analog and digital voice to circuit and packet data. The interface between users of the HF service and the HF system is a *subnetwork interface* conceptually similar to that of STANAG 5066.

A *session manager* coordinates the components of the 3G HF system to provide the requested service. It conceptually provides a scheduling service, arbitrating among service requests from higher-layer protocols. The queuing algorithm may involve priority, time to live, and any backoff algorithms that deal with network congestion and fairness of service. Although this function is a key element in the

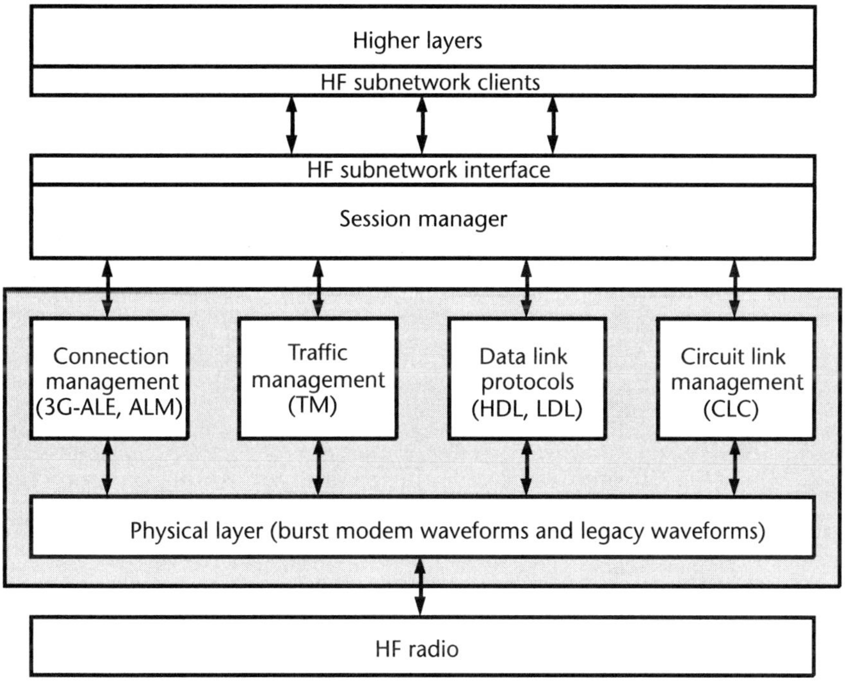

Figure 5.1 The 3G HF technology suite.

effectiveness of a 3G system, the session managers at communicating HF stations do not directly communicate with each other; their specifications were therefore judged not to affect interoperability, so the session management function is not standardized.

The standardized 3G protocols and waveforms are shown surrounded by a gray box in Figure 5.1:

- The *connection management* function sets up, maintains, and tears down the HF links as required for the requested communication services. The new, more efficient 3G ALE is included here (also called link setup or LSU), along with a new automatic link maintenance (ALM) function.
- *Traffic management (TM)* coordinates traffic flow and which of the available communication protocols will be used on a link after it is set up. In some cases, TM can be accomplished during link setup.
- A new suite of packet-oriented data link protocols is introduced in the 3G suite: the high-throughput data link (HDL) and low-latency data link (LDL).
- For circuit-oriented applications (both analog and digital), the *circuit link management* function is engaged after link setup to coordinate use of the link in circuit mode.
- A family of PSK burst waveforms supports these functions. This family is scalable over a wide range of robustness to the challenges of the HF skywave channel.

The 3G suite was designed to operate with the same HF radio technology as the older ALE and data modems[1].

5.2 Burst Waveforms

Many services provided by the 3G HF system require relatively short transmissions:

- Channel probing;
- Link setup, link maintenance, negotiation of traffic protocols, and network time synchronization;
- Acknowledging receipt of data.

Burst waveforms can be very efficient and robust for conveying such short messages, and the 3G system uses them extensively. For simplicity of specification, the standard also treats the waveforms of longer and more varied duration used for data transfer (BW2 and BW3) as burst waveforms.

This section discusses the range of burst waveforms used to meet the distinctive requirements of the various 3G protocols as to payload, duration, time synchronization, and acquisition and demodulation performance in the presence of noise, fading, and multipath. We begin with an overview of the scalable burst waveform structure, followed by detailed descriptions of the specific bursts used by the 3G protocols. Further discussion of this scalable burst waveform family may be found in [2].

5.2.1 Generic Structure of the Burst Waveforms

The generic burst consists of three distinct sections, a transmit level control (TLC) section, an acquisition preamble, and a data section. Each of these sections consists of multiple frames of a PN-spread 8-ary PSK waveform. Each frame comprises 32, 64, or 96 8-ary PSK symbols, which are generated at a rate of 2400 symbols per second. After filtering, the PSK frames quadrature modulate an 1800-Hz carrier to provide a noise-like, transmit signal with a 3-kHz bandwidth.

5.2.1.1 TLC Section

Existing HF radios were generally not designed with burst waveforms in mind. For example, MIL-STD-188-141 military radios are allowed 25 ms to reach full transmit power after keying. While the transmitter radio frequency stages are ramping up, the input audio signal level is adjusted by a transmit level control (TLC) loop so that it fully modulates the transmit power. At the receiver, an automatic gain control (AGC) loop must also adjust to a new receive signal. To accommodate these characteristics of existing radios, the 3G burst waveforms begin with a TLC section

1. As in the previous generation, the timing characteristics of the radio, antenna coupler, and so on affect the operation of the protocols. Programmable timing parameters are provided to accommodate this, and must be set uniformly across the network to ensure interoperability.

of "throwaway" 8-ary PSK symbols that are passed through the system while the transmitter's and receiver's level control loops stabilize.

5.2.1.2 Preamble

The 3G burst waveform preamble is composed of a number of 96 symbol (40 ms) 8-ary PSK frames. The length of the preamble varies among the 3G bursts to balance the speed versus acquisition robustness requirements of the various applications. In general, longer preambles provide increased performance, especially in fading conditions where shorter preambles may be missed, at the expense of increased transmit duration and receive processing.

5.2.1.3 Data Section

The structure of the data section of 3G bursts is determined by the error correction coding scheme selected, the interleaving of the encoded data bits, and the modulation technique.

Error Correction Coding

After convolutional codes, BCH codes, and Hadamard codes were investigated, a convolutional encoding scheme with soft-decision Viterbi decoding was selected due to its superior performance and ability to scale to the exact number of payload data bits required [3]. Convolutional coding techniques have been popular choices for HF applications when a continuous data stream of many bits is to be transmitted (e.g., the modems in Chapter 3). However, for the very short payloads in a burst (as few as 26 bits), the straightforward application of a convolutional code can be inefficient. A rate $1/n$ constraint length K convolutional coder is typically flushed by the encoding of $K-1$ known bits (usually zeroes) immediately after the last user bit. These known tail bits can represent a significant portion of the on-air energy when the number of user bits is small. For example, a rate 1/2 , $K = 7$ convolutional encoding scheme used with a 26-bit payload would spend 6/32 or nearly 20% of its energy in sending to these tail bits. To reduce this overhead a simple tail-biting technique [4] is implemented in many of the 3G burst waveforms: the encoder's state memory is initialized with the final $K-1$ user payload bits, instead of $K-1$ bits known to the receiver.

Interleaving

The coded bits of the Data section are interleaved using a rectangular block interleaver[2]. For each burst waveform, the number of coded payload bits is used to select dimensions of the "most square-shaped" array that contains that number of bits. Bits are written into the interleaver row-wise and read from the interleaver column-wise for transmission, with this process reversed at the receiver. Even over the rela-

2. BW3 is an exception: it uses an interleaver structure in which (in general) both the row and column indices are changed between successive insertions into the interleaver. The standard refers to this as a *convolutional block interleaver structure*.

tively short durations of the 3G bursts, this interleaving is useful for spreading out bursts of errors to permit more effective error correction at the receiver.

Modulation

The narrowband data waveforms described in Chapter 3 employ an orthogonal Walsh function based waveform for the most robust applications, and equalized PSK or QAM for higher data rates. Likewise, most 3G burst waveforms use Walsh coding for robustness. An equalized PSK waveform is used only for sending data packets with the high-throughput data link protocol.

5.2.1.4 *3G Burst Waveform Characteristics*

Table 5.1 summarizes the characteristics of the 3G burst waveforms and their uses. These bursts are described in more detail in the following sections. The protocols that use these bursts to carry their protocol data units (PDUs) are presented in Sections 5.3 through 5.6.

5.2.2 Burst Waveform 0 (BW0)

The short, robust burst waveform 0 (BW0) burst used by the robust link setup (RLSU) protocol (see Section 5.3.5) exemplifies much of the philosophy of the 3G burst waveform family. The burst begins (see Figure 5.2) with 256 "throwaway" symbols that are sent while the transmitter level control and receiver AGC are settling. This is followed by a preamble containing 384 PSK symbols of known data. This acquisition preamble provides an opportunity for the receiver to detect the presence of the waveform and to estimate various parameters for use in data demodulation. The remainder of the burst carries a payload of 26 protocol bits.

The 26 protocol bits are encoded using a $r = 1/2$, $k = 7$ convolutional encoder (Figure 5.3) with tail biting. The 52 coded bits are then interleaved using a 4×13 block interleaver. Four bits at a time are fetched from the interleaver, and these are used to select one of 16 orthogonal Walsh functions (see Table 5.2) to be sent in the data portion of the burst. Each 16-symbol sequence from Table 5.2 is sent four times, resulting in a 64-symbol sequence on the air for every 2 bits of payload. The Walsh sequences are added modulo-8 to a pseudo-noise (PN) sequence for transmission.

We can get a sense of the robustness of this process by noting that each on-air, 8-PSK symbol could carry 3 bits of information; we are effectively using 96 bits of information capacity on the channel to convey each bit of payload. This is the computation reflected in the "Effective Code Rate" column in Table 5.1.

5.2.3 Burst Waveform 1 (BW1)

Burst waveform 1 (BW1) is a general-purpose vehicle used to carry short messages for many of the 3G protocols: traffic management, link maintenance, and data acknowledgments for the HDL protocol.

The BW1 burst (Figure 5.4) is a longer, more robust version of BW0. A longer preamble (240 ms) provides improved acquisition performance. A 48-bit payload

Table 5.1 Burst Waveform Characteristics

Wave Form	*Used for*	*Burst Duration1*	*Payload*	*Preamble*	*FEC Coding*	*Interleaving*	*Data Format*	*Effective Code Rate2*
BW0	Robust LSU PDUs	613.33 ms 1472 PSK symbols	26 bits	160.00 ms 384 PSK symbols	Rate = 1/2, *k* = 7 convolutional (no flush bits)	4x13 block	16-ary Walsh function	1/96
BW1	Traffic management PDUs; HDL Ack PDUs	1.30667 seconds 3136 PSK symbols	48 bits	240.00 ms 576 PSK symbols	rate = 1/3, *k* = 9 convolutional (no flush bits)	16x9 block	16-ary Walsh function	1/144
BW2	HDL traffic data PDUs	126.67 + (*n**400) ms 304 + (*n**960) PSK symbols, n = 3, 6, 12, or 24	n*1881 bits	26.67 ms 64 PSK symbols (for equaliser training)	rate = 1/4, *k* = 8 convolutional (7 flush bits)	None	32 unknown/ 16 known	variable: 1/1, 1/2, 1/3, 1/4, ...
BW3	LDL traffic data PDUs	373.33 + (*n**13.33) ms	8*n*+25 bits	266.67 ms 640 PSK symbols	rate = 1/2, *k* = 7 convolutional (7 flush bits)3	convolutional block	16-ary Walsh function	variable: 1/12, 1/24, ...
BW4	LDL acknowledg-ment PDUs	640.00 ms	2 bits	None	None	None	4-ary Walsh function	1/1920
BW5	Fast LSU PDUs	1.01333 seconds 2432 PSK symbols	50 bits	240.00 ms	Rate = 1/2, *k* = 7 convolutional (no flush bits)	10x10 block	16-ary Walsh function	1/96

1. All transmissions begin with a TLC/AGC guard sequence (part of the preamble for BW3). These symbols are included in the indicated burst durations.
2. Reflects forward error correction (FEC) and Walsh-function coding only, relative to uncoded 8-PSK; does not include known data or convolutional encoder flush bits.
3. In this case, the number of flush bits exceeds by one the minimum number required to flush the convolutional encoder; this makes the number of coded bits a multiple of four as is required for the Walsh-function modulation format.

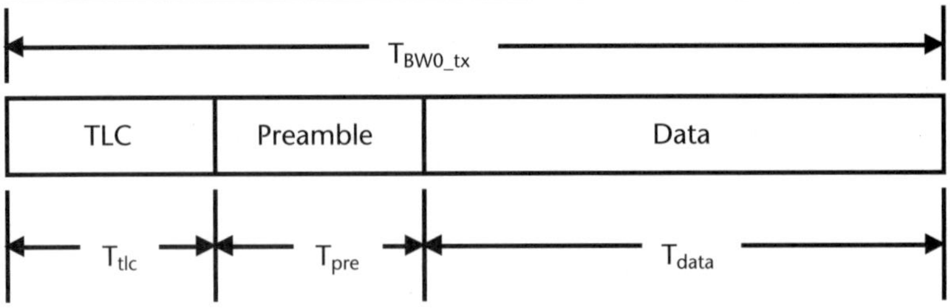

T_{tlc} = 106.667 ms: 256 PSK symbols at 2400 symbols/second

T_{pre} = 160.000 ms: 384 PSK symbols at 2400 symbols/second

T_{data} = 346.667 ms: 832 PSK symbols at 2400 symbols/second (interleaved and PN spread)

T_{BW0_tx} = total duration = 613.333 ms

Figure 5.2 BW0 structure.

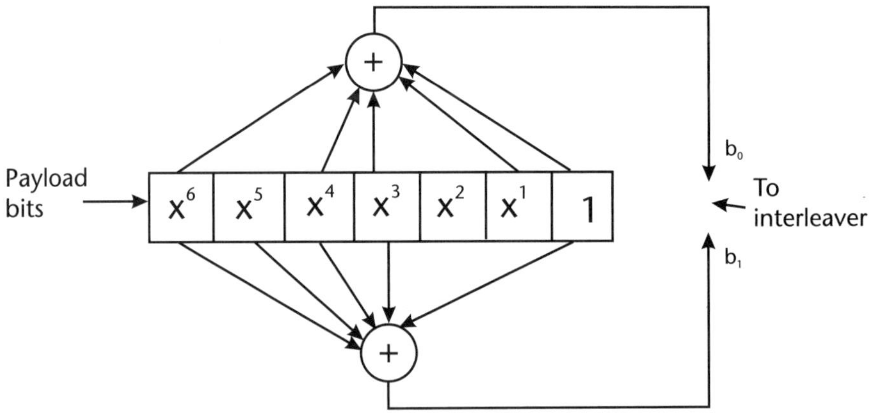

Figure 5.3 BW0 convolutional coder.

carries more information in a single burst, and also requires a larger interleaver so BW1 has additional time diversity and increased robustness to fading; a $r = 1/3$, $k = 9$ convolutional coder (Figure 5.5) provides increased immunity to noise. The 16-ary Walsh sequences are repeated four times (identical to BW0).

5.2.4 Burst Waveform 2 (BW2)

Burst waveform 2 (BW2) is qualitatively different from the other 3G burst waveforms. It is used by the HDL protocol to carry user data, and was designed for speed rather than robustness. Following a 100 ms TLC section and a short (26.67 ms), PN-spread preamble, BW2 contains a negotiated number NumPKT of fixed-size data packets (Figure 5.6). NumPKT, one of 3, 6, 12, or 24, is negotiated before the HDL protocol begins, and remains unchanged until the end of the data transfer. The modulation used in the data section of a BW2 burst is a 4800 bps equalized waveform similar to that described in Chapter 3, rather than the orthogonal Walsh sequences used in all of the other burst waveforms. Each data packet comprises 20

Table 5.2 Walsh Modulation of Coded Bits to Tribit Sequences

Coded Bits	*Tribit Sequence*
0000	0000 0000 0000 0000
0001	0404 0404 0404 0404
0010	0044 0044 0044 0044
0011	0440 0440 0440 0440
0100	0000 4444 0000 4444
0101	0404 4040 0404 4040
0110	0044 4400 0044 4400
0111	0440 4004 0440 4004
1000	0000 0000 4444 4444
1001	0404 0404 4040 4040
1010	0044 0044 4400 4400
1011	0440 0440 4004 4004
1100	0000 4444 4444 0000
1101	0404 4040 4040 0404
1110	0044 4400 4400 0044
1111	0440 4004 4004 0440

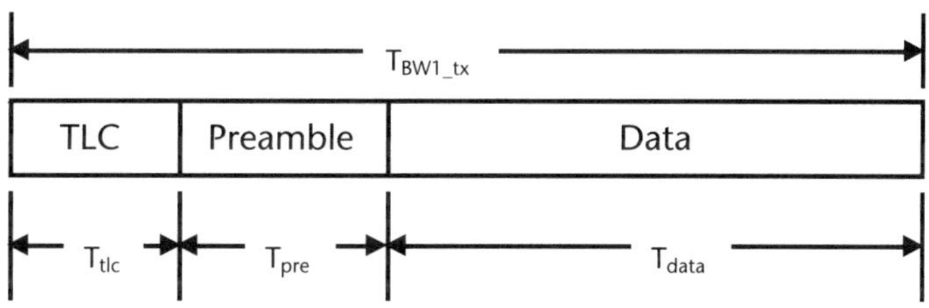

T_{tlc} = 106.667 ms: 256 PSK symbols at 2400 sym/second
T_{pre} = 240.000 ms: 576 PSK symbols at 2400 sym/second
T_{data} = 960.000 ms: 2304 PSK symbols at 2400 sym/second (interleaved and PN spread)

T_{BW1_tx} = total duration = 1306.667 ms

Figure 5.4 BW1 structure.

frames, each of which contains 32 8-PSK symbols of coded payload data, followed by 16 known symbols of probe data. Tail biting is not used in BW2; each data packet contains 1913 payload bits that are passed through the convolutional coder, plus 7 flush bits.

The 3G data link protocol that uses BW2 bursts, HDL, is a Type II hybrid ARQ protocol, wherein data is initially sent at an FEC coding rate $r = 1$. That is, the number of bits sent in the initial transmission of a packet equals the number of payload bits in that packet. If the receiver cannot successfully decode the packet

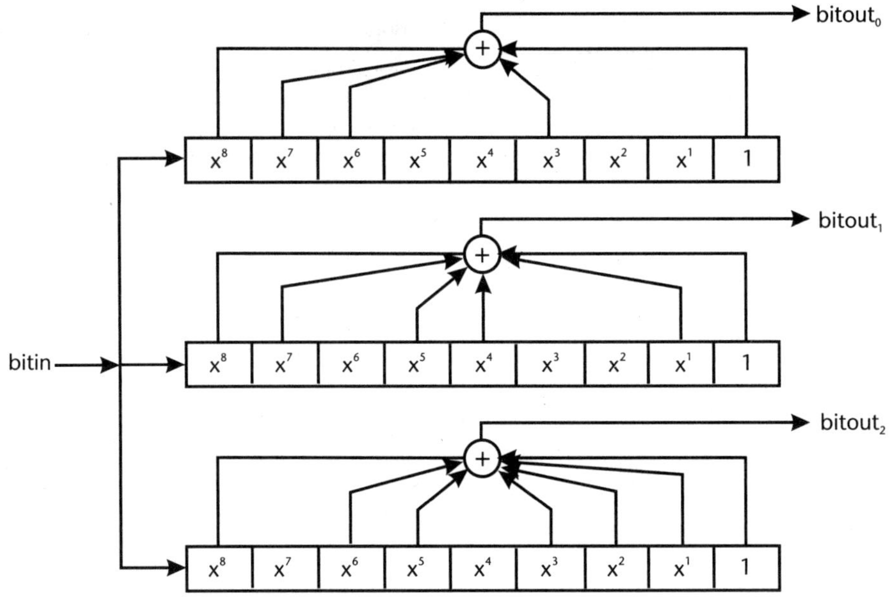

Figure 5.5 BW1 convolutional coder.

(indicated by CRC failure), it requests retransmission; each retransmission carries *additional* FEC bits. Thus, FEC redundancy is added only as needed.

The BW2 FEC is a rate r = 1/4, k = 8 convolutional code (Figure 5.7). The initial transmission of a packet consists of the $Bitout_0$ sequence shown in Figure 5.7; retransmissions carry the remaining three Bitout sequences in rotation. A fifth transmission of the same packet would repeat the $Bitout_0$ sequence, except that the 3-bit symbols are rotated ($M_2M_1M_0$ becomes $M_0M_2M_1$) a different number of times for each transmission. Soft decision code combining is used to decode BW2 packets, as described in the HDL Section 5.5.3.

5.2.5 Burst Waveform 3 (BW3)

Burst waveform 3 (BW3) is a robust burst used for carrying data packets in the LDL protocol. Each burst begins with a combined TLC and preamble (see Figure 5.8), followed by a single data packet. The length of the data packet is negotiated prior to starting the LDL protocol, and is then fixed until the end of the data transmission. This length of the BW3 data section can be any multiple of 32 bytes from 32 through 512 bytes.

BW3 uses the same rate r = 1/2, k = 7 convolutional FEC as BW0, but without tail biting; seven flush bits are appended to each sequence of coded payload bits[3]. Like HDL, LDL is also a Type II hybrid ARQ protocol. In the first transmission of a packet, the $Bitout_0$ sequence is interleaved, Walsh-coded, PN-spread, and sent over the air. Retransmissions, if needed, alternate between $Bitout_1$ and $Bitout_0$. Unlike

3. Since BW3 uses a k = 7 convolutional code, only six bits are needed to flush the encoder. The seventh flush bit is added purely for convenience (to make the number of coded bits per BW3 transmission a multiple of four), so that each group of four bits can then be mapped to an orthogonal Walsh symbol.

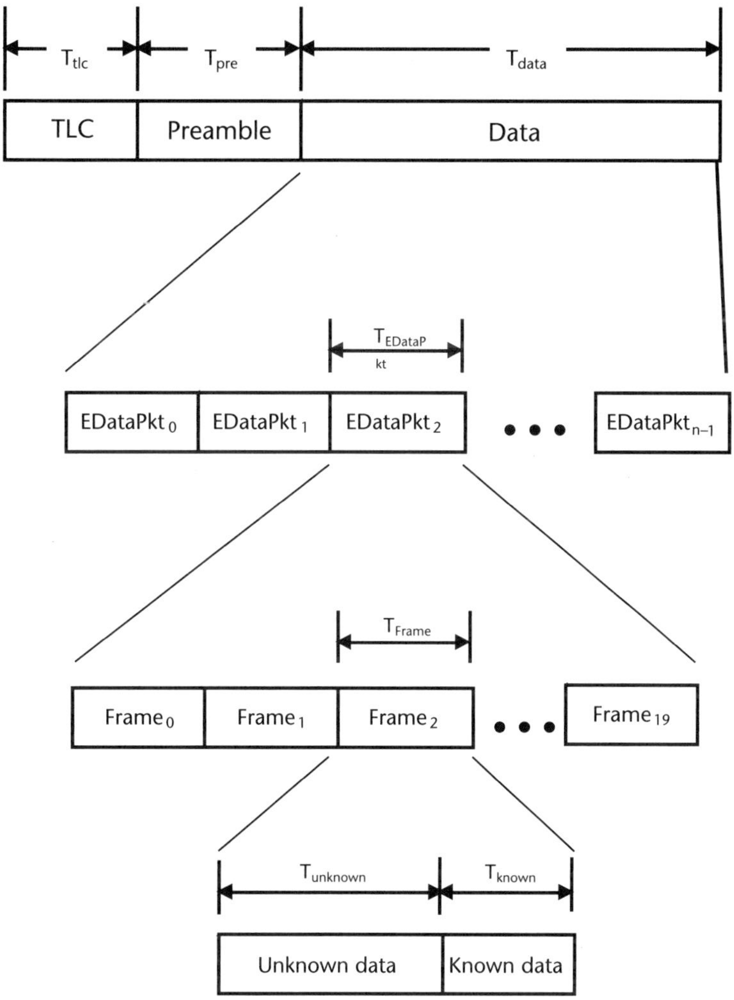

Figure 5.6 BW2 structure.

the other Walsh-coded 3G bursts, each 16-symbol Walsh sequence in a BW3 data burst is sent only once (not four times).

5.2.6 Burst Waveform 4 (BW4)

Burst waveform 4 (BW4) is a very robust burst used for carrying acknowledgment packets in the LDL protocol. The BW4 payload is only two bits: an ACK/NAK bit and an EOM flag. As usual, each burst begins with a TLC section, but no preamble is sent (see Figure 5.9). Instead, the two payload bits are used directly (no FEC or interleaving) to select one of four 16-symbol orthogonal Walsh sequences. The selected Walsh sequence is then sent 80 times, with PN spreading.

5.2.7 Burst Waveform 5 (BW5)

Burst waveform 5 (BW5), used by the FLSU and FTM protocols, is an extended version of the BW0 (RLSU) burst (see Figure 5.10). It uses the same TLC, FEC, Walsh

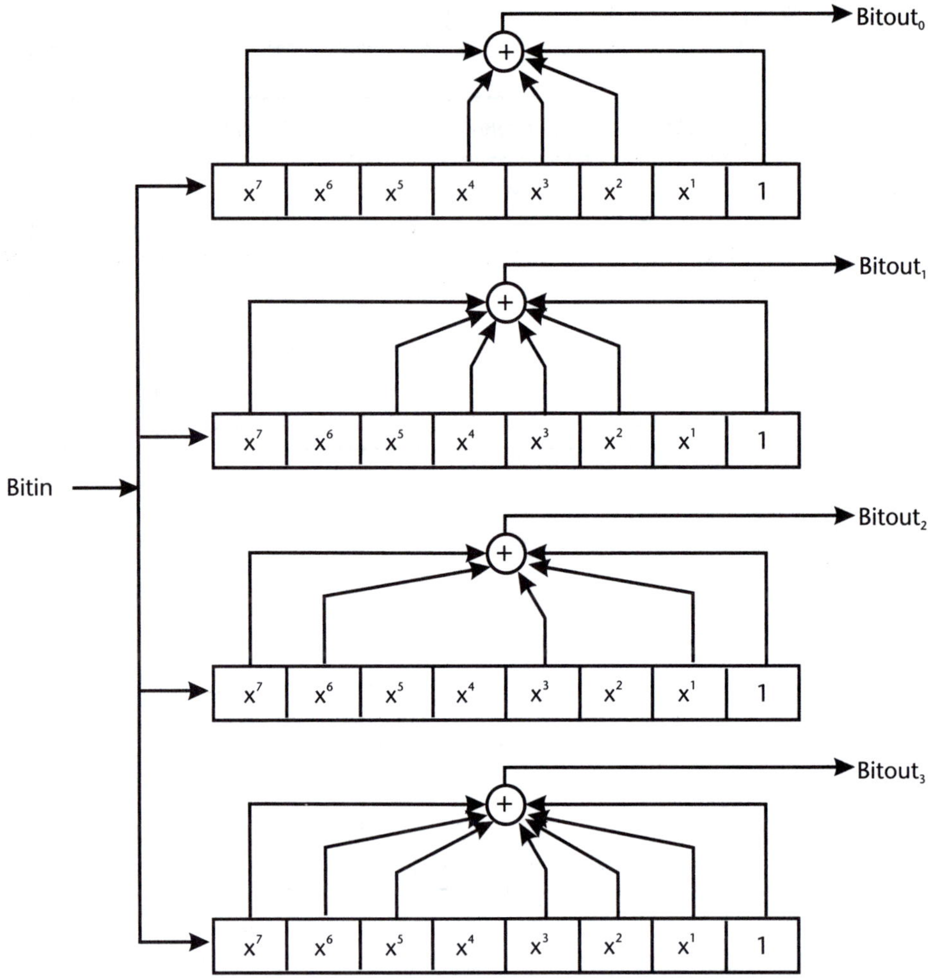

Figure 5.7 BW2 convolutional coder.

coding, and PN spreading as BW0. A longer preamble and a 50-bit payload (with increased interleaver span) make BW5 more robust than BW0.

5.3 Third-Generation Automatic Link Establishment

The first of the 3G protocols that we will discuss is automatic link establishment (ALE). 3G ALE is designed to quickly and efficiently establish one-to-one and one-to-many (both broadcast and multicast) links. Compared to 2G ALE, the 3G approach is faster, more efficient, and more robust to the challenges of the HF channel. This results from several key advances:

- The burst waveforms used for 3G ALE (BW0 and BW5) are roughly 10 dB more robust than the 8-ary FSK waveform used for 2G ALE.

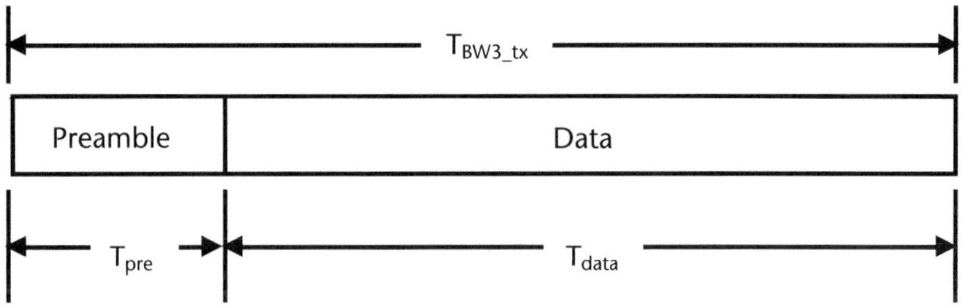

T_{pre} = 266.67 ms: 640 PSK symbols at 2400 symbols/second

T_{data} = 106.67 + (n * 13.33) ms: 32n + 256 PSK symbols (interleaved and PN spread)

where n = 32 * m bytes, and m = 1, 2, 3, 4, ..., 16

T_{BW3_tx}= 373.33 + (n*13.33) ms (ranges from 0.8 through 7.2 s)

Figure 5.8 BW3 structure.

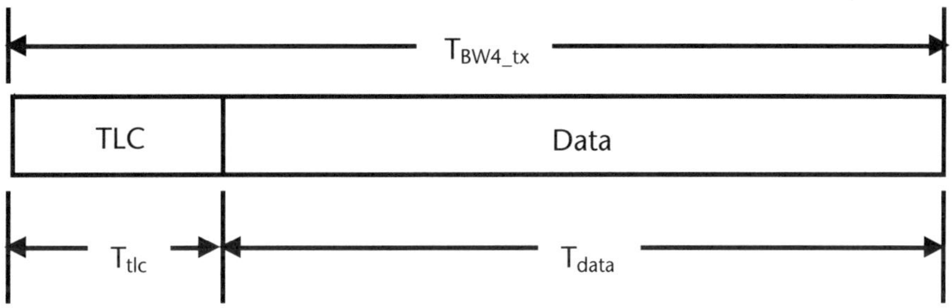

T_{tlc} = 106.67 ms: 256 PSK symbols at 2400 sym/second

T_{data} = 533.33 ms: 1280 PSK symbols at 2400 sym/second

T_{BW4_tx}= 640.00 ms

Figure 5.9 BW4 structure.

- 3G ALE normally operates in synchronous mode, which eliminates the need for the long scanning call required by the asynchronous 2G ALE system.
- 3G addresses are fixed-size, and are about half as long as the shortest 2G address. This results in shorter calling PDUs, and therefore faster calls.
- The notion of *trunking* is introduced, wherein channels used for setting up links can be separated from channels used for traffic. This can improve overall network efficiency.
- Various anticollision mechanisms are used in the link setup (LSU) protocols to reduce the rate of calls failing due to collisions.

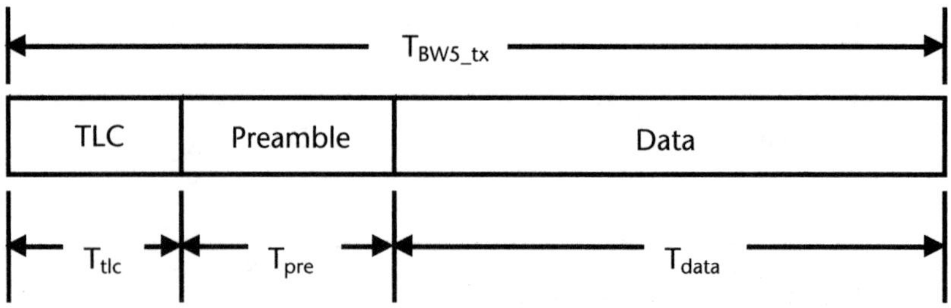

T_{tlc} = 106.667 ms: 256 PSK symbols at 2400 sym/second

T_{pre} = 240.000 ms: 576 PSK symbols at 2400 sym/second

T_{data} = 666.667 ms: 1600 PSK symbols at 2400 sym/second (interleaved and PN spread)

T_{BW5_tx} = total duration = 1013.333 ms

Figure 5.10 BW5 structure.

3G ALE includes two distinct LSU protocols: fast link setup (FLSU) and robust link setup (RLSU). FLSU is optimized for speed in setting up links in small networks, while RLSU is designed to perform well in large networks and under heavy traffic loads. These protocols are not interoperable, and only one should be in use in a 3G HF network.

This section begins with a discussion of the aspects of 3G ALE that are common to the two protocols, and then discusses FLSU and RLSU individually.

5.3.1 Synchronous Operation

As in the previous generations of ALE systems, all available 3G HF receivers continuously scan an assigned list of calling channels, listening for 2G or 3G calls[4]. However, 2G ALE is an asynchronous system in the sense that a calling station makes no assumption about when a destination station will be listening to any particular channel. The 3G HF system includes a similar asynchronous mode; however, synchronous operation is the preferred mode for 3G networks, as it will usually provide better speed of linking[5] and efficiency of spectrum use.

When operating in synchronous mode, all scanning receivers in a 3G ALE network change frequency at the same time, to within a relatively small timing uncertainty (see Figure 5.11). The current dwell channel of every station in a network can always be computed from the time of day and the address of the desired station. Thus, a synchronous 3G call doesn't require the extended 2G ALE scanning call to capture a scanning receiver; instead, a very short call on the known dwell channel will suffice to reach a receiver that is not linked on some other channel.

4. All 3G systems are required to recognize incoming 2G calls, and to respond using the 2G protocol (unless responses are specifically inhibited). This ensures backward interoperability.

5. It is conceivable that a network in synchronous operation could take *longer* to establish a link if the time to step synchronously to a usable channel is longer than the duration of an asynchronous call (which could begin immediately on the desired channel).

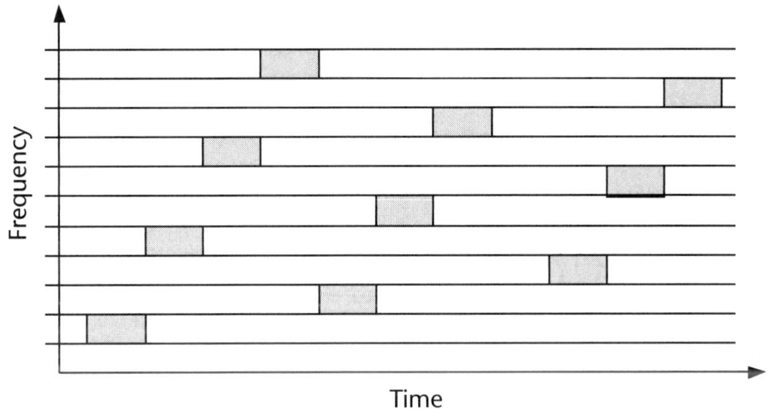

Figure 5.11 Synchronous scanning.

Note that it is not necessary that all stations monitor the same calling channel at the same time. By assigning groups of network members to monitor different channels in each scanning dwell (Figure 5.12), simultaneous calls directed to different member stations will be distributed in time or frequency, which greatly reduces the probability of collisions among 3G ALE calls. This is especially important under high-traffic conditions. The set of stations that monitor the same channels at the same time is called a *dwell group*.

Of course, in some applications it is valuable for all stations in a network to be in the same dwell group so that they overhear calls to each other. They can thereby track when other stations will be linked and when they will be unavailable for calls.

Synchronous scanning permits rapid completion of a call, since no scanning call is required. However, when a link is required to use a specific frequency (e.g., if only that frequency is propagating), an asynchronous system may link faster. This is because the synchronous system must wait for the called receiver to dwell on the

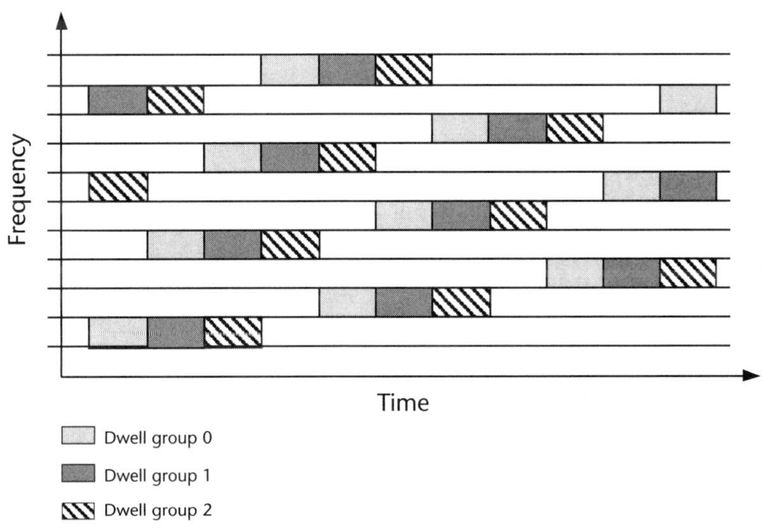

Figure 5.12 Dwell groups.

desired frequency, and this requires, on average, waiting through half of the scan cycle.

5.3.2 3G Frequency Management

The automatic channel selection (ACS) function of 3G ALE is invoked to select channels for calling and traffic. A well-conceived ACS function would be aware of various attributes for each channel in the pool, such as the IONCAP predicted SNR, recent sounding information, recent occupancy information, and latency until the channel is next scanned.

3G networks can operate using a single pool of frequencies for both calling and traffic (as in previous generations of HF ALE). However, large networks *with a large pool of frequencies* can make more efficient use of the spectrum by segregating channels used for calling from those used for traffic.

Calling channels are used in an unscheduled manner to make contact with other stations. Ideally, calling channels would always be vacant, and therefore available for immediate use by any station that needs to set up a link. However, traffic channels are engaged after the parties to a link are already in contact, and can therefore be managed to achieve very high utilization.

Thus, large 3G networks with heavy traffic and a large number of channels available will generally operate more effectively in such a trunking mode (i.e., with separate calling and traffic channels). Smaller or lightly loaded networks, on the other hand, may forego the complexity of trunked operation and employ a small pool of channels for both calling and traffic, just as in previous generations of ALE.

5.3.3 3G-ALE Addressing

One of the functions of the subnetwork layer in Figure 5.1 is translation of upper-layer addresses (e.g., IP addresses) into the addressing scheme used in the local subnet. The addresses used in 3G-ALE PDUs are 10-bit binary numbers. For comparison, even the shortest 3-character addresses used in 2G ALE provide over 15 bits of name space. Although the 3G name space is much smaller than that available with 2G ALE, 3G networks may nevertheless have up to 1000 stations, which even the U.S. DoD agreed was sufficient for their largest contemplated HF networks. Also note that while the user-friendly ASCII call signs used as station addresses in 2G ALE may still be used in a 3G network—they must now be translated by the equipment to and from the small binary 3G addresses used over the air.

In NATO, HF stations are addressed using a 13-bit network number and a 10-bit address within that network. The latter field is called the point/multipoint address because it may refer either to a single station or to a collection of stations sharing a single address. The 3G ALE PDUs naturally accommodate the 10-bit point/multipoint addresses of the NATO addressing structure. The 13-bit network number is used as follows to ensure that cross-network addressing ambiguity is avoided:

- The network number of the called station (or collective) is used in the linking protection (LP) applied to the 3G ALE PDU, as shown in Figure 5.13. It is replicated to match the length of the LP encryption key in use in the network,

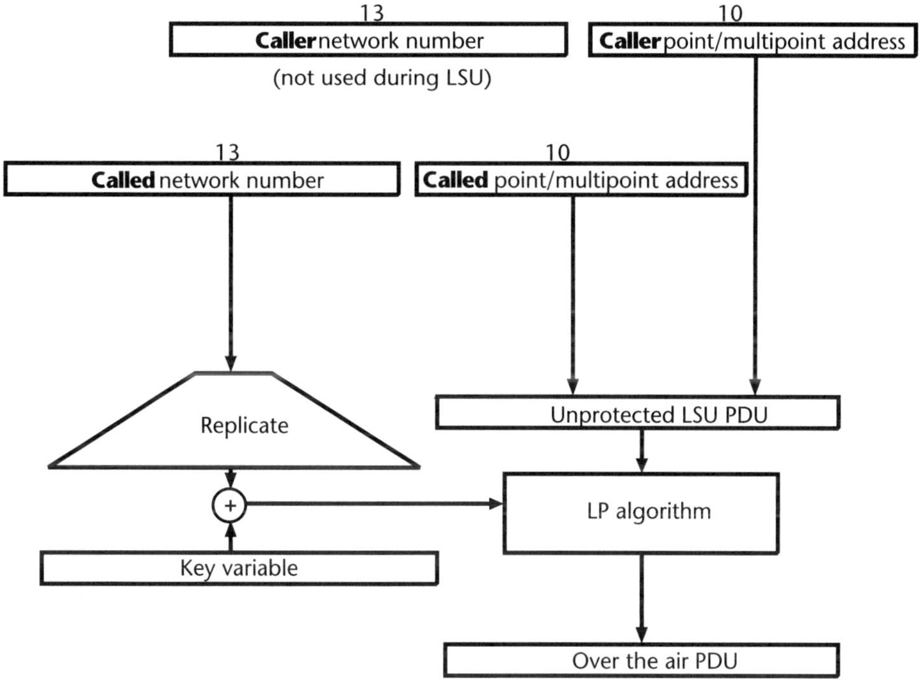

Figure 5.13 NATO-mode addressing.

and then exclusive-ored with that key for use as the key in the LP algorithm. Stations that receive this protected PDU will attempt to decrypt it using their local network number(s). If the network number is not the same as that of the called station, the decryption will fail, and the receiver will ignore the PDU. This scheme thus ensures that the network number is used to disallow links to unintended networks just as effectively as if the network number bits were sent explicitly.

- The network number of the *calling* station is not used during link setup. The same scheme cannot be used to mix the calling network number into the over-the-air PDU because the called station(s) do not know a priori which networks may call them. Any necessary authentication of the caller is therefore deferred until after link setup.

5.3.4 Fast Link Setup

This section describes the fast link setup (FLSU) protocol and its closely associated FTM protocol, which are sometimes referred to together as simply FLSU. As FLSU is setting up a link, it also conveys the traffic type that will be used immediately after the link setup is complete. Thus, FLSU accomplishes both LSU and the initial traffic management negotiation. FTM is most often used after completing one data transfer to set up a subsequent transfer (including reversing the traffic direction on the link when clients are exchanging packets in both directions).

Fast link setup offers a range of capabilities:

- Point-to-point (PTP) link setup, analogous to a 2G ALE individual call.
- Point-to-multipoint link (PTM) setup, analogous to a 2G ALE net call.
- Link termination, analogous to use of the 2G ALE TWAS word.
- Time distribution, analogous to the 2G ALE capability, but with the increased precision needed to support synchronous operation in the absence of external time sources such as GPS. Time of day (TOD) uncertainty at each FLSU station must not exceed 184.16667 ms for synchronous operation.

The basic exchange in FLSU is a two-way handshake[6]. In the example PTP link setup in Figure 5.14, the caller sends a *request* PDU and the called station responds with a *confirm* PDU. The figure also shows some key timing parameters of FLSU, including the 1.35 s dwell time. We illustrate the capabilities of FLSU in more detail following a description of the FLSU PDUs.

5.3.4.1 FLSU Protocol Data Units

The PDUs used by FLSU and FTM are carried in the 50 bits of a BW5 burst. PDUs for the two protocols are distinguished by the protocol field in the first three bits: 001 for an FLSU PDU versus 100 for an FTM PDU (see Figure 5.15).

Priority

In all REQ and some CONF PDUs (i.e., FTM or FLSU PDUs having Type = REQUEST or CONFIRM), the next two bits indicate the priority of the traffic to be carried on the link, when established; 0 indicates the highest priority traffic, and 3 the lowest. Traffic priority is also used in the FLSU backoff algorithm, described

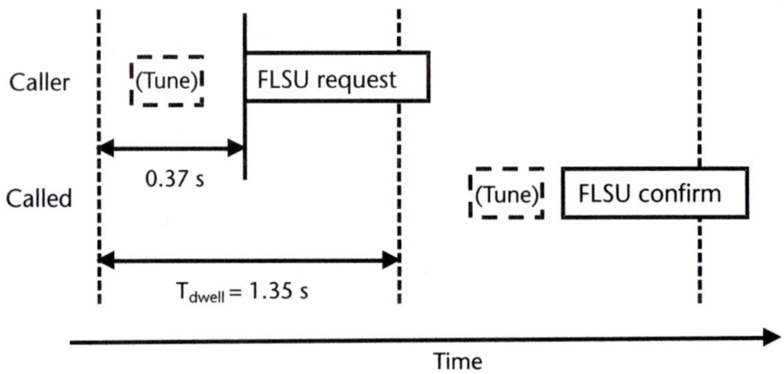

Figure 5.14 FLSU two-way handshake.

6. By comparison, 2G ALE includes a third transmission to complete the handshake. This is to confirm to the called station that the link is active. In 3G ALE, this confirmation is found in the startup of traffic (or the traffic management protocol) from the calling station, so only a two-way handshake is required in the LSU protocols. In FLSU in particular, a calling station that has failed to receive an error-free response PDU transmits a very robust link termination PDU. The robustness of the link termination PDU allows the called station to receive it very reliably if it has been transmitted. Hence, if the called station does not receive a link termination PDU, this indicates with high probability that the link has been successfully established.

bits	3	2	10	10	1	1	3	6	6	8
FLSU	001	Pri	Dest Addr	Src Addr	Addr Type	XN	PDU Type	Arg 1	Arg 2	CRC
FTM	100	Pri	Dest Addr	Src Addr	Addr Type	XN	PDU Type	Arg 1	Arg 2	CRC

Figure 5.15 FLSU protocol data units.

in Section 5.3.4.2. In CONF PDUs, this field is used only when the argument field value refers to one of the high-rate ('HDL_n') or LDL ('LDL_n') traffic types, as shown in Table 5.5. In TOD_Response PDUs (Type = 5), the field-value is set to LOW (3) if the time reference being transmitted is based on locally-received GPS time; otherwise, it is set to ROUTINE (2). The field-value is set to 3 (LOW) in all other CONF PDUs and in all TERM PDUs, and must be ignored by the receiver.

Addresses

The Dest Addr field carries the address of the station (or stations) to which this PDU is being sent. Dest Addr can be the individual address of a single intended recipient station, a multicast address, or all ones for a broadcast link. When the destination address is a multicast address or a broadcast address, the Addr Type field is set to "1". In nonlinking calls (specifically, LQA sounds), Dest Addr is set to the address of the sending station; hence it has the same value as Source Addr.

The Source Addr field contains the address of the station that is sending this PDU. It is always the station address of a single station—never a multicast or broadcast address.

The XN field indicates that the destination is in the same network as the caller when set to 0. When set to 1, the destination is in a different network, and a station in the local network with the same 10-bit address must not respond.

PDU Type

The PDU Type field indicates the role of the PDU in the protocol. Encodings of this field are shown in Table 5.3.

Argument Fields

The two argument fields in the PDU carry information specific to the protocol and PDU type, as indicated in Table 5.4. For example, link setup requests carry the channel to be used for traffic and the traffic type (see Table 5.5) in these fields.

CRC Field

The CRC field contains an 8-bit cyclic redundancy check (CRC) for error detection, computed over the preceding 42 bits of the PDU. The generator polynomial is $X^8 + X^7 + X^4 + X^3 + X + 1$.

Table 5.3 FLSU PDU Type

Type	*Description*
0	REQUEST_2Way: a "request with acknowledgment" shown in Figure 5.14
1	CONFIRM: confirms the sender's readiness for the requested service.
2	TERM: terminates the sender's participation in the current service.
3	Asynchronous FLSU_REQUEST: sent multiple times, followed by a single FLSU request
4	REQUEST_1Way: a "request without acknowledgment"
5	TOD_Response: response to a REQ_2Way whose traffic type = TOD. Carries the TOD minutes (0…59) in Argument1 and TOD seconds (0..59) in Argument2.
6..7	reserved

Table 5.4 FLSU Argument Field Usage

Field	*Values*	*Description*	
Argument1	PDU Type	FLSU	FTM
	0	Channel (0..63)	Duration (TBD)
	1	TOD Correction / Score	Score
	2	Channel (0..63) or Score in LQA Exchange	Reserved
	3	Channel (0..63)	Reserved
	4	Channel (0..63)	Duration (TBD)
	5	Minutes (0..59)	Reserved
	6..7	Reserved	Reserved
Argument2	PDU Type	FLSU	FTM
	0	Traffic Type (Table 5.5)	Traffic Type (Table 5.5)
	1	TOD Correction / Score	Score
	2	Reason code	Reason code
	3	Traffic Type (Table 5.5)	Reserved
	4	Traffic Type (Table 5.5)	Traffic Type (Table 5.5)
	5	Seconds (0..59)	Reserved
	6..7	Reserved	Reserved

5.3.4.2 FLSU Protocol Operation

When a link is required, the ACS function is first invoked to select the calling channel for the linking attempt. A service primitive is then issued to the FLSU function specifying a single datagram, traffic mode, and the channel upon which to link. The FLSU function then attempts to link on the specified channel and simply indicates success or failure to the session manager. In the event of a failed linking attempt, the session manager is responsible for issuing another transmission request service primitive for the datagram at a later time.

Table 5.5 Traffic Types (FLSU and RLSU)

Code	*Traffic Type*
0	NO_TRAFFIC_TO_SEND
1	ANLG_VOICE
2	DGTL_VOICE
3	ANDVT [S-4197, S-4198] (parameters are autodetected)
4	S-4285 [2400, long intlv]
5	S-4285 [2400, short intlv]
6	S-4285 [1200, long intlv]
7	S-4285 [1200, short intlv]
8	S-4285 [600, long intlv]
9	S-4285 [600, short intlv]
10	S-4285 [300, long intlv]
11	S-4285 [300, short intlv]
12	S-4285 [150, long intlv]
13	S-4285 [150, short intlv]
14	S-4285 [75, long intlv]
15	S-4285 [75, short intlv]
16	S-4415 (autodetect of long/short intlvr)
17	S-4539_HDR (parameters are autodetected)
18	SER_110B (parameters are autodetected)
19	HDL_24
20	HDL_12
21	HDL_6
22	HDL_3
23 .. 38	LDL_32, 64, 96, ... , 512
39	LQA
40	HDL+
41 ... 50	Reserved for future S-4538 use
51 ... 62	Reserved for vendor use (noninteroperable)
63	TOD (FLSU only)

FLSU *by itself* is a best effort protocol for *single* linking attempts, rather than a persistent protocol that tries multiple times on multiple frequencies. Of course, a frequency selection (ACS) algorithm and persistence protocols are required in a system, but they do not affect on-air interoperability and are not standardized.

FLSU cstablishcs logical links using either a one way PDU or a two-way PDU exchange. A link termination, as a third transmission, must be sent if the call response is not received to avoid half-up links.

Collision Avoidance Algorithm

A linking failure is detected at the calling station when the response expected from the called station is garbled or missing. Such failures may be due to poor propagation, a blocked channel at the receiver, a collision with another transmission, and so on. FLSU attempts to avoid collisions by invoking a backoff algorithm upon any linking failure. When the failure is detected, the calling station must wait a randomly selected number of dwell times before trying again. The range of backoff times depends on the priority of the traffic. Suggested ranges of backoff times as a function of traffic priority are shown in Table 5.6.

FLSU Examples

Figures 5.16 through 5.22 describe (by means of specific example scenarios) some of the capabilities provided by the FLSU protocol. The following scenarios are presented:

- Synchronous two-way LSU, point-to-point packet service;
- Synchronous two-way LSU, point-to-point circuit mode service;
- Synchronous two-way LSU failure, assuming that the point-to-point packet service was requested;
- Asynchronous two-way LSU, point-to-point packet service;
- Synchronous two-way net LSU, circuit mode service (conference mode);
- Time of day (TOD) distribution via HF means;
- Synchronous two-way LSU, point-to-point packet service, showing optional trunked operation with separate calling and traffic channels.

All the scenarios show a two-dimensional (time and frequency) view, with 4 or 8 frequencies listed on the horizontal axis, and time on the vertical axis (time progresses from top to bottom). Unless otherwise indicated, calling frequencies and traffic frequencies are common (identical). A legend depicts how stations are identified in the figures:

- Light gray depicts the caller activities;
- White depicts the called station activities;
- Dark gray (cross-hatch) depicts the activities of all the net members.

Synchronous Two-Way FLSU, Point-to-Point Packet Service

Beginning at the top left corner, Figure 5.16 shows that all stations in the net synchronously scan the assigned frequencies. The dwell time is 1.35 seconds per

Table 5.6 FLSU Backoff Times

Priority	*Backoff (dwells)*
Highest	1-2
High	1-4
Routine	1-8
Low	1-16

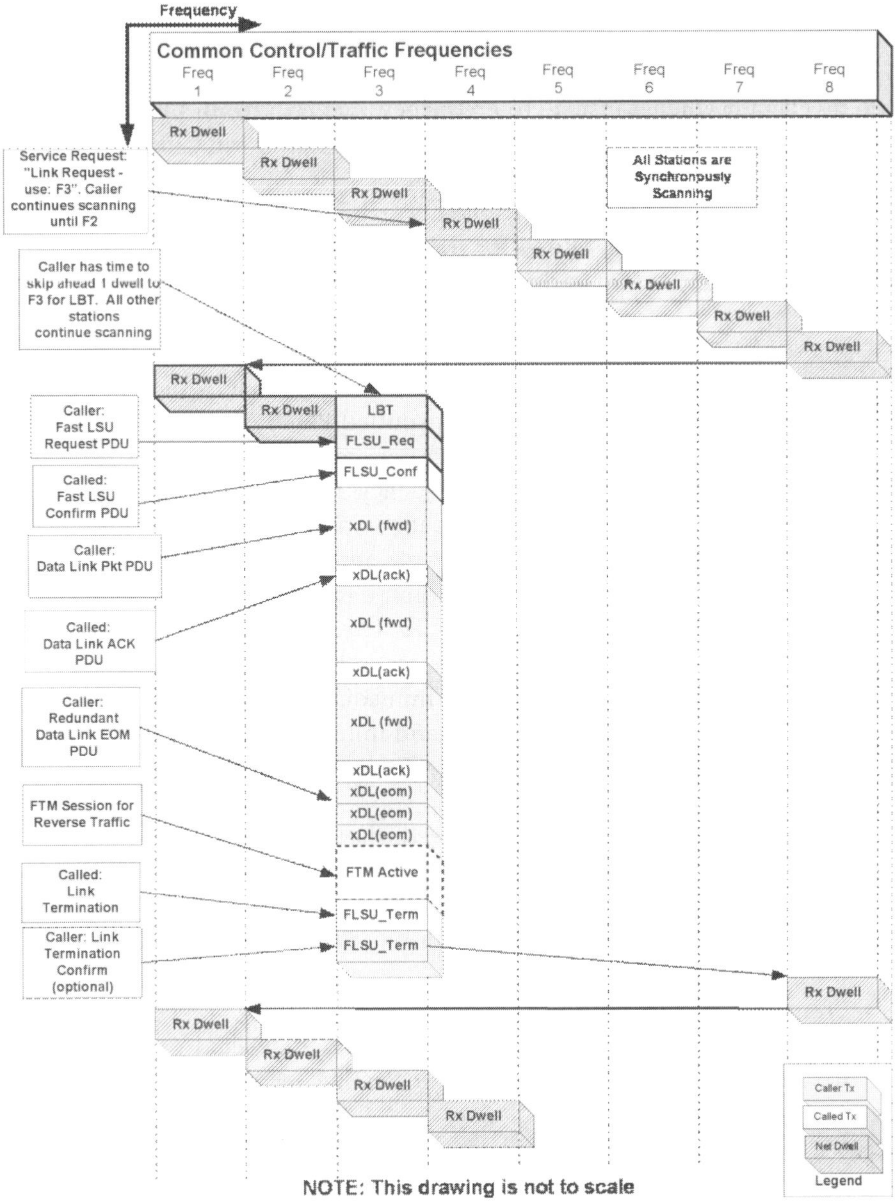

Figure 5.16 Synchronous two-way FLSU, point-to-point packet service.

frequency. While scanning, all the stations are required to perform the listen before transmit (LBT) algorithm as a means of establishing a frequency occupancy perspective for each of the frequencies in the scan list. During the dwell on Frequency 4, a station is directed to establish a point-to-point link, with a specific station, on Frequency 3 (F3), using the xDL (generic reference to either HDL or LDL) ARQ protocol for reliable packet transfer.

The caller station continues scanning until one dwell prior to desired calling frequency. During this period, the caller is still available to respond to any incoming higher- and equal-priority calls. If this takes place, then the original intended call is deferred.

Otherwise, at the end of the period the caller station skips the Frequency 2 dwell and switches instead to Frequency 3, executing an LBT process to assure that the channel is unoccupied. The remaining stations will continue scanning synchronously until they come upon Frequency 3. Note that if the service request specifying F3 was issued just prior to the normal F3 dwell (such that LBT is not possible), then the specification allows for transmitting on F3 if the occupancy data acquired previously during the normal scanning process is deemed reliable.

During the Frequency 3 time slot, the caller station issues a two-way FLSU_Request PDU, which conveys the caller station address, called station address, priority, and the desired traffic service (xDL ARQ mode). All stations in the net will stop scanning if they detect the transmitted PDU. All stations except for the called station are free to resume scanning after determining that the call is not for them.

The called station stays on Frequency 3 and responds with an FLSU_Confirm PDU, indicating the ability to continue with the requested traffic service. Both the caller and called stations then enter the agreed xDL protocol; they alternate sending xDL PDUs, with the caller sending data using the xDL_DATA PDU, and the called responding with the xDL ACK/NAK PDU. This process continues until all data has been transferred error-free, as indicated by the caller sending redundant xDL end of message (EOM) PDUs. Immediately after the xDL transfer is complete, both stations remain linked on F3 and initiate the fast traffic management (FTM) protocol to negotiate further traffic. This gives the called station an opportunity to send reverse traffic.

After a link timeout has occurred, the last station to receive an xDL transfer terminates the link by sending an FLSU_Term PDU. After terminating the link, both the caller and called rejoin the other net members in synchronous scanning. Note that in this scenario, the caller station performed a lengthy LBT prior to transmitting the FLSU_Req PDU. If the link request had occurred just prior to the Frequency 3 dwell, the LBT process would not have taken place because the station had been executing LBTs on each scanning dwell already and had presumably determined that Frequency 3 was unoccupied.

Synchronous Two-Way FLSU, Point-to-Point Circuit Mode Service

The scenario in Figure 5.17 is identical to the above scenario with the exception that the traffic service is *circuit mode*. The FLSU_Request specifies the traffic waveforms that will be used during circuit mode. For example, STANAG 4285 can be specified as the traffic waveform. Once circuit mode begins, any station can initiate transmissions using the specified traffic waveform. A CSMA/CA process is recommended to avoid collisions.

Synchronous Two-Way FLSU Failure: Packet Traffic Example

Figure 5.18 shows the required procedure for a failed link setup. All two-way FLSU calls require only a request and confirm PDU transmission. A third transmission is issued only if the caller station does not correctly receive a confirm PDU as expected.

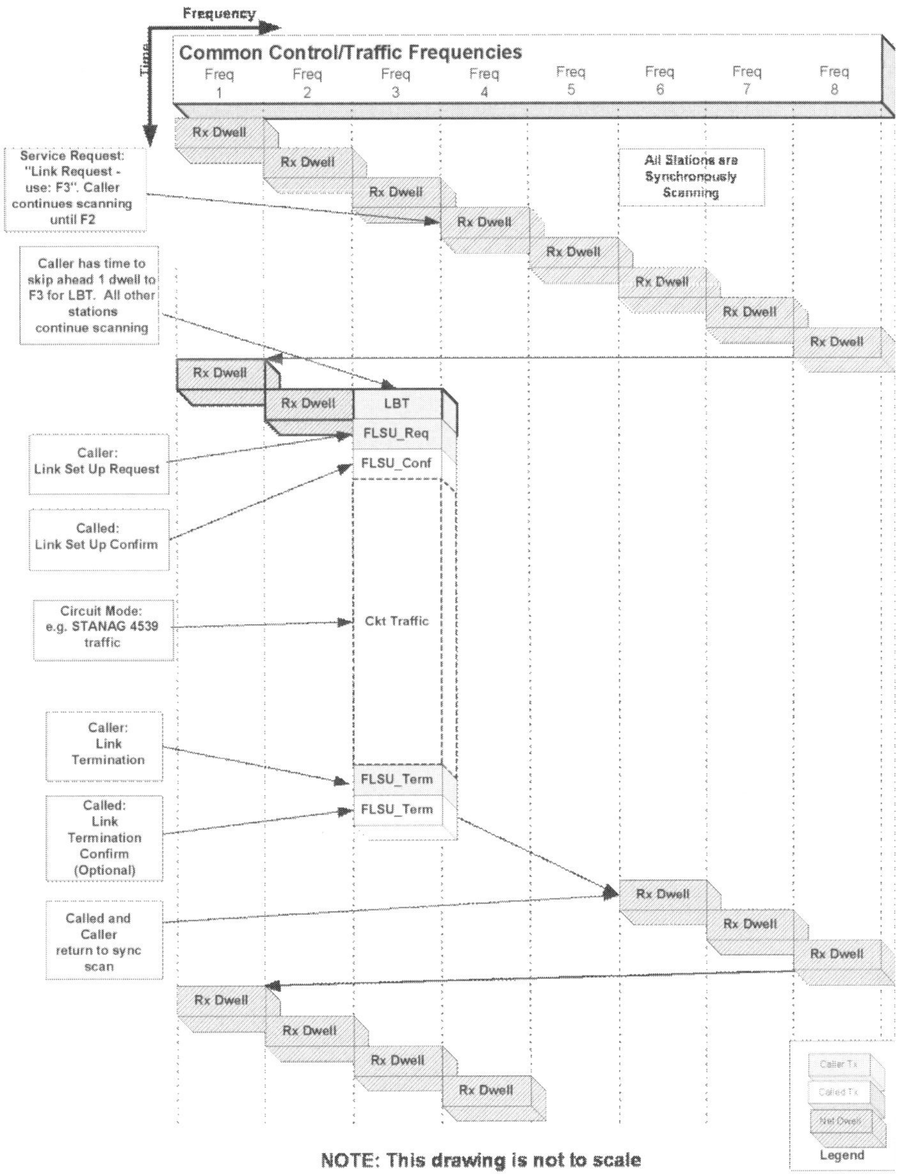

Figure 5.17 Synchronous two-way FLSU, point-to-point circuit mode service.

There can be many reasons for such a failure to link, such as CRC failure, propagation failure, an unexpected result in any field of the FLSU_Confirm PDU, or reception of an unexpected PDU of a different type. In these cases, the caller station is required to transmit an FLSU_Terminate PDU.

However, the caller must honor the requirement that a receiving station need not execute more than dual demodulation. The scenario shown depicts the case in which the xDL ARQ protocol is invoked via the original FLSU call. Since the calling station did not receive the FLSU_Confirm response, it must assume that a response was issued but that it did not propagate correctly, and that the called station

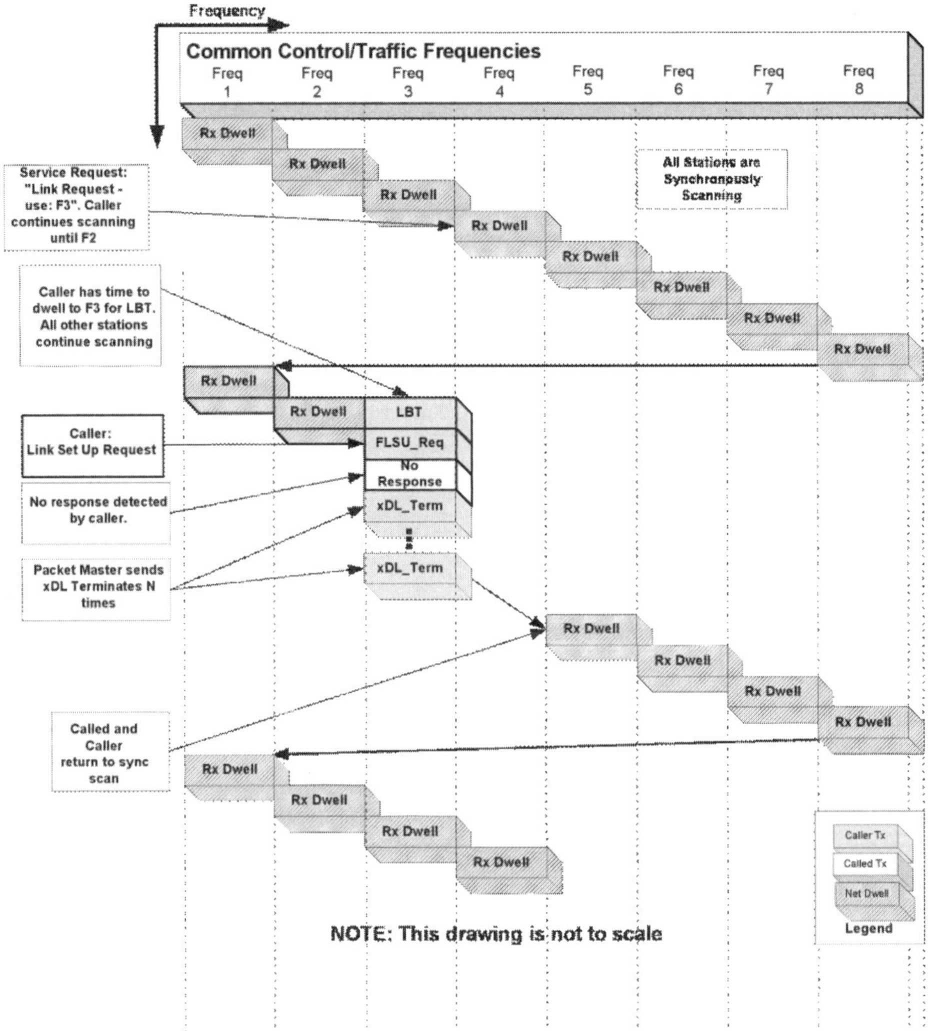

Figure 5.18 Synchronous two-way FLSU failure.

is prepared for the xDL packet transfer protocol. As such, the called station is set up to receive either the first xDL forward packet PDU, or an xDL_Terminate PDU. Sending an FLSU_Terminate would impose a triple demodulation requirement on the receiving station. Thus, the calling station must send up to *N* "xDL_Terminate" PDUs to abort the ARQ protocol. Under the xDL protocol specification, *N* is defined by the number of xDL_Terminate PDUs that would fit within the time slot of a forward packet PDU. If this were a circuit traffic example, the "xDL_Terminate" PDUs would not be necessary, and the calling station could send the FLSU_Terminate PDU immediately after the failed call response.

Asynchronous Two-Way FLSU, Point-to-Point Packet Service

A station without net synchronization can still initiate a call using the asynchronous calling procedure. The call type (point-to-point, point-to-multipoint, etc.) and the traffic service types are identical to those allowed during a synchronous call, as shown in Figure 5.19.

An unsynchronized calling station scans the allocated frequencies using the required dwell rate. However, it is assumed that it is not scanning synchronously relative to the other net members. The asynchronous call begins with the LBT (for at least one dwell period), followed by the transmission of about 1.35*N* Async_FLSU_Request PDUs on the requested link frequency, where *N* is the number of channels in the scan list, and 1.35 is the duration of each dwell (in seconds). Transmitting 1.35*N* Async_FLSU_Request PDUs guarantees that all other scanning stations will scan the calling channel during the async call, even under the worst-case time of day offset conditions. If the time of day offset can be estimated more

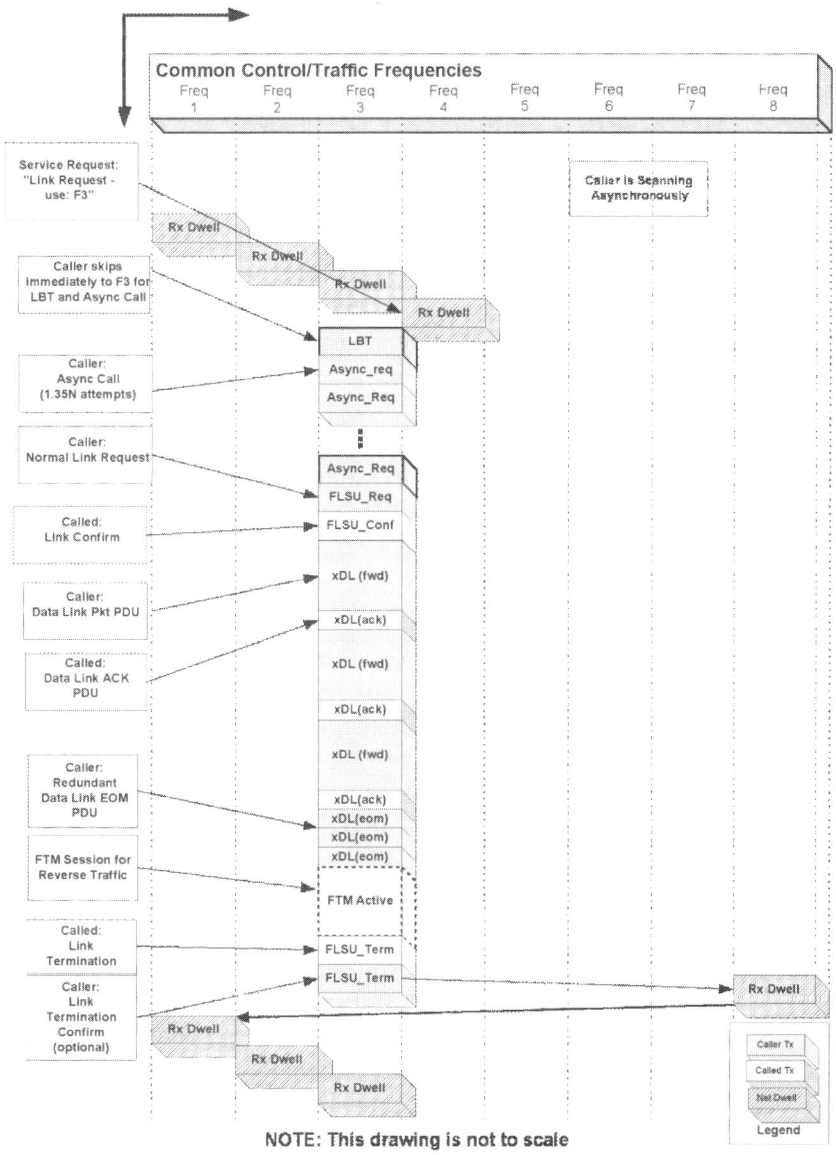

Figure 5.19 Asynchronous two-way FLSU.

accurately, fewer than 1.35N Async_FLSU_Request PDUs may be sent to capture the desired station.

Since the address of the called station(s) is contained in the Async_FLSU_Req PDU, all stations that are not included in the call are free to resume scanning. Called station(s) that receive one of the asynchronous FLSU PDUs stop scanning and wait for the normal FLSU_Request PDU, which is sent immediately after the final Async_FLSU_Request PDU. The maximum wait duration is approximately equal to $1.35(N + 1)$ seconds, where N is the number of frequencies in the scan list.

After receiving a valid FLSU_Request PDU, the addressed station responds normally with the FLSU_Confirm PDU. All subsequent elements of the FLSU protocol are identical to the synchronous case. The BW5 FLSU burst waveform (and the required dwell timing) have been designed to assure reception of the asynchronous call (given an open frequency and adequate propagation conditions).

Synchronous Two-Way, Point-to-Multipoint FLSU, Circuit Mode

The scenario in Figure 5.20 is identical to the PTP circuit mode scenario presented above, except that it is a point-to-multipoint (PTM) call. Within the two-way FLSU_Request, the called station address is a multicast address (addresses a group of stations within the network). This type of call (two-way) demands that the called stations respond sequentially (a roll-call, similar to the slotted responses to a 2G ALE net call) in an order specified by their station address. One can see that each station responds with an FLSU_Confirm PDU during its allocated time slot. As in the above scenario, the traffic type portion of the FLSU_Request PDU specifies the traffic waveform.

Note that if a *one-way* point-to-multipoint (PTM) FLSU_Request PDU were used by the caller, the called stations would not respond. A roll-call response is only used if a two-way call is selected by the caller.

Any station can issue an FLSU_Terminate (link) PDU, announcing its departure from the link. If the caller station issues a sequence of FLSU_Terminate PDUs using the multicast address, all stations should return to scan mode (this may follow a confirmation of link termination by each station, if invoked).

It is possible that some stations miss a multicast call, either due to a temporary propagation anomaly, or because they were linked on a different frequency during the call. The caller station can reissue the FLSU_Request PDU to the multicast address on a different frequency, selected to capture net members that missed the original call. The roll-call process would be repeated.

Active Time of Day (TOD) Distribution via FLSU

The diagram in Figure 5.21 shows the procedure for TOD distribution, transferring TOD to an unsynchronized calling station. The unsynchronized station transmits 1.35N Asynchronous FLSU_Req PDUs using the asynchronous calling technique described previously. The FLSU_ Request PDU (with traffic type set to TOD) is transmitted once, at the end of the asynchronous calling period. The destination address may be all 1's, an implicit address which indicates that the reigning net control station should be the only responder, or the (explicit) address of any station. After transmitting the TOD request, the requesting station monitors the calling channel

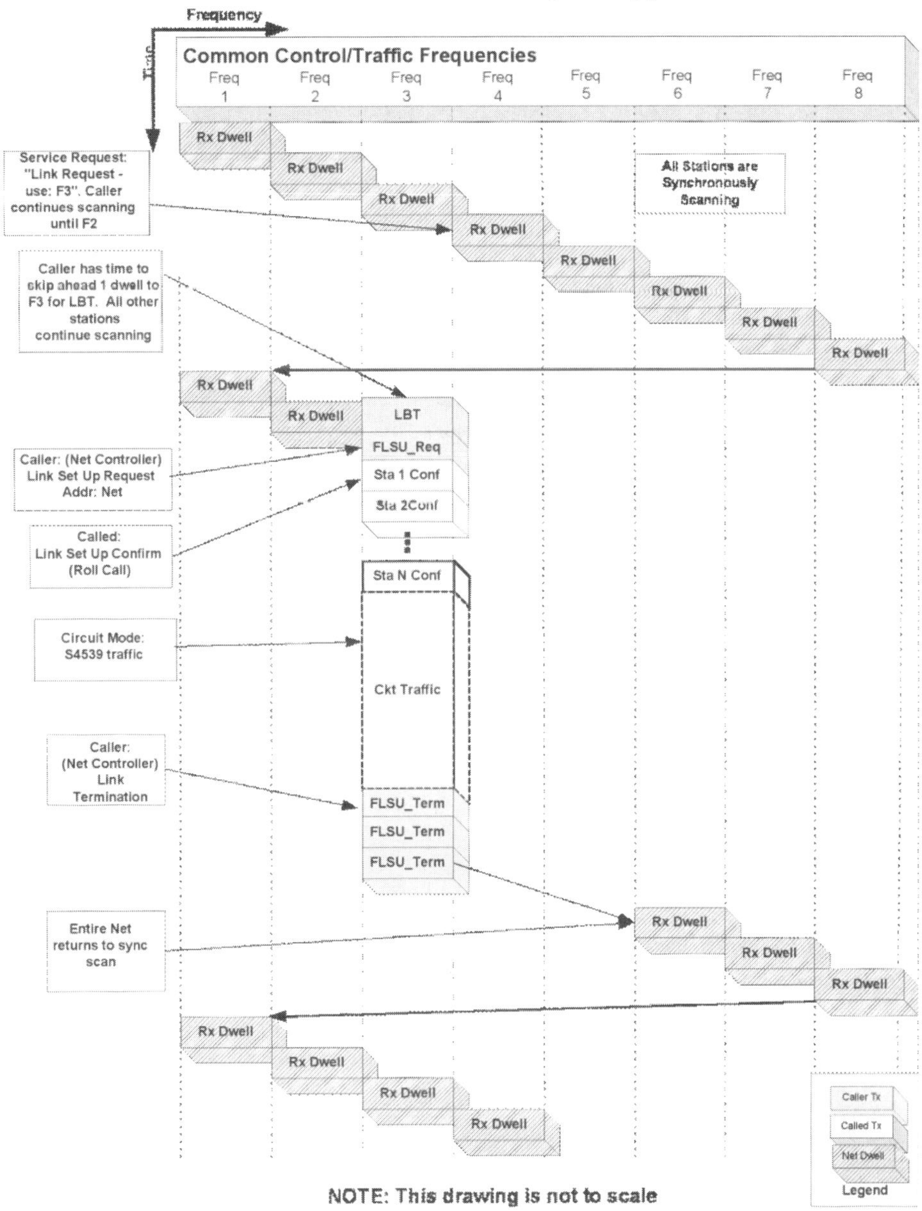

Figure 5.20 Synchronous two-way, point-to-multipoint, FLSU, circuit mode.

for the TOD_Response PDU. The monitoring timeout period is defined as two scanning dwell periods.

After receiving a TOD request, the (explicitly or implicitly) addressed station transmits the FLSU_TOD_Response PDU on the original calling channel. The TOD_Response PDU transmission must meet the precise timing requirements of a synchronized PDU transmission. The TOD_Response PDU contains the relevant TOD information. In case of a CRC failure on the TOD_Response PDU, the calling station must repeat the entire process.

There are other methods for TOD synchronization:

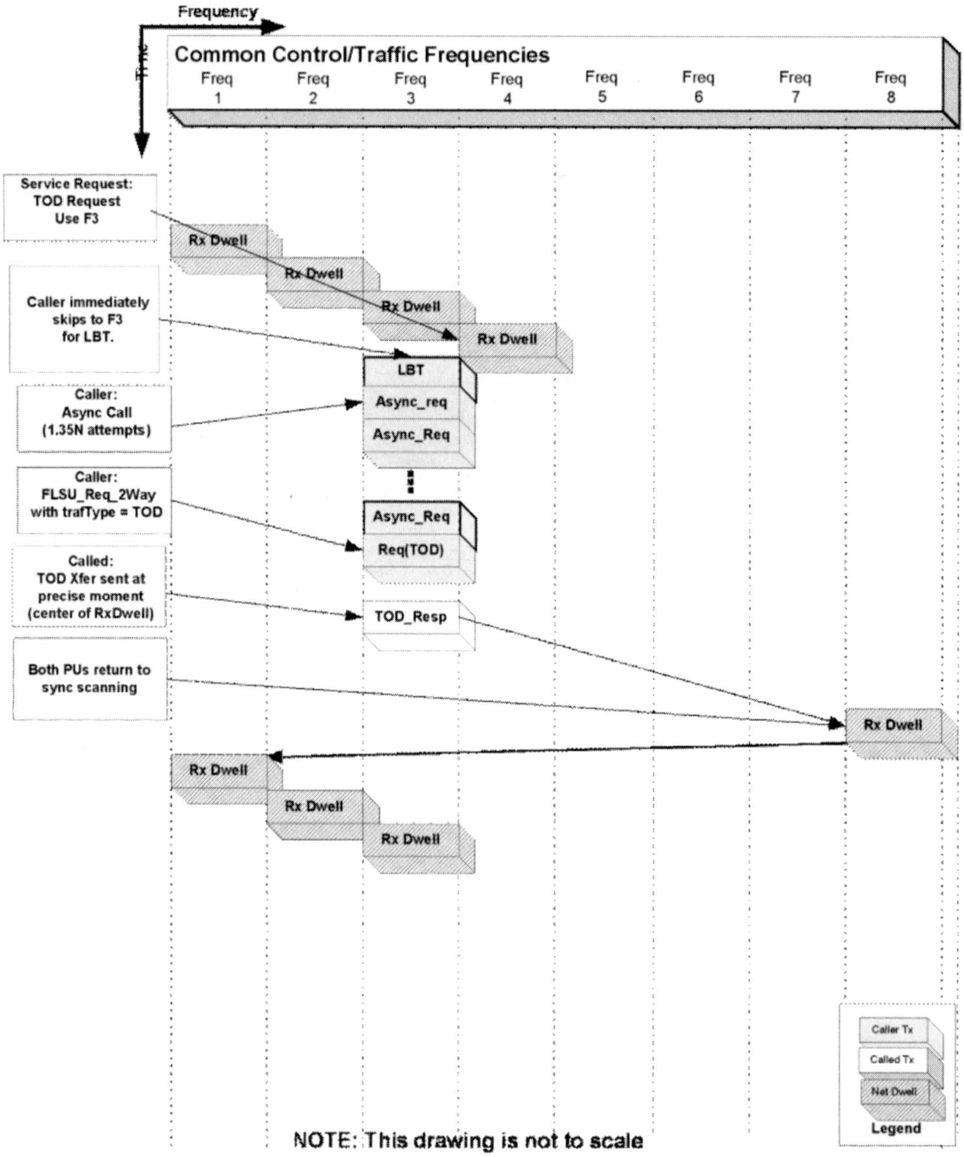

Figure 5.21 Active TOD distribution via FLSU.

- *GPS TOD sync* is the preferred method, since it is both passive and extremely accurate (but it relies on an external service).
- *Passive TOD sync* can be established by searching a specific channel within the scan list for synchronized calls (but the accuracy of TOD acquired in this manner is usually not known, and further complicated by linking protection).
- Lastly, *broadcast TOD distribution* can be achieved by the net control station issuing both the TOD_Request and TOD_Response PDUs. All stations that monitor the broadcast TOD can then receive the TOD sync passively.

Synchronous Two-Way FLSU, Point-to-Point Packet Service, with Separate Calling/Traffic Channels

The diagram in Figure 5.22 describes the optional capability of using separate calling and traffic channels. Normally, FLSU uses the calling channel for traffic as well. However, the PDUs and required timing parameters fully support using separate calling and traffic channels. The example scenario is similar to previous synchronous calling scenarios, except that an additional dwell is introduced for the purpose of assessing occupancy on the desired traffic channel.

Initially, all stations in this scenario are scanning synchronously. The calling station receives a request from its user process for a link establishment using calling

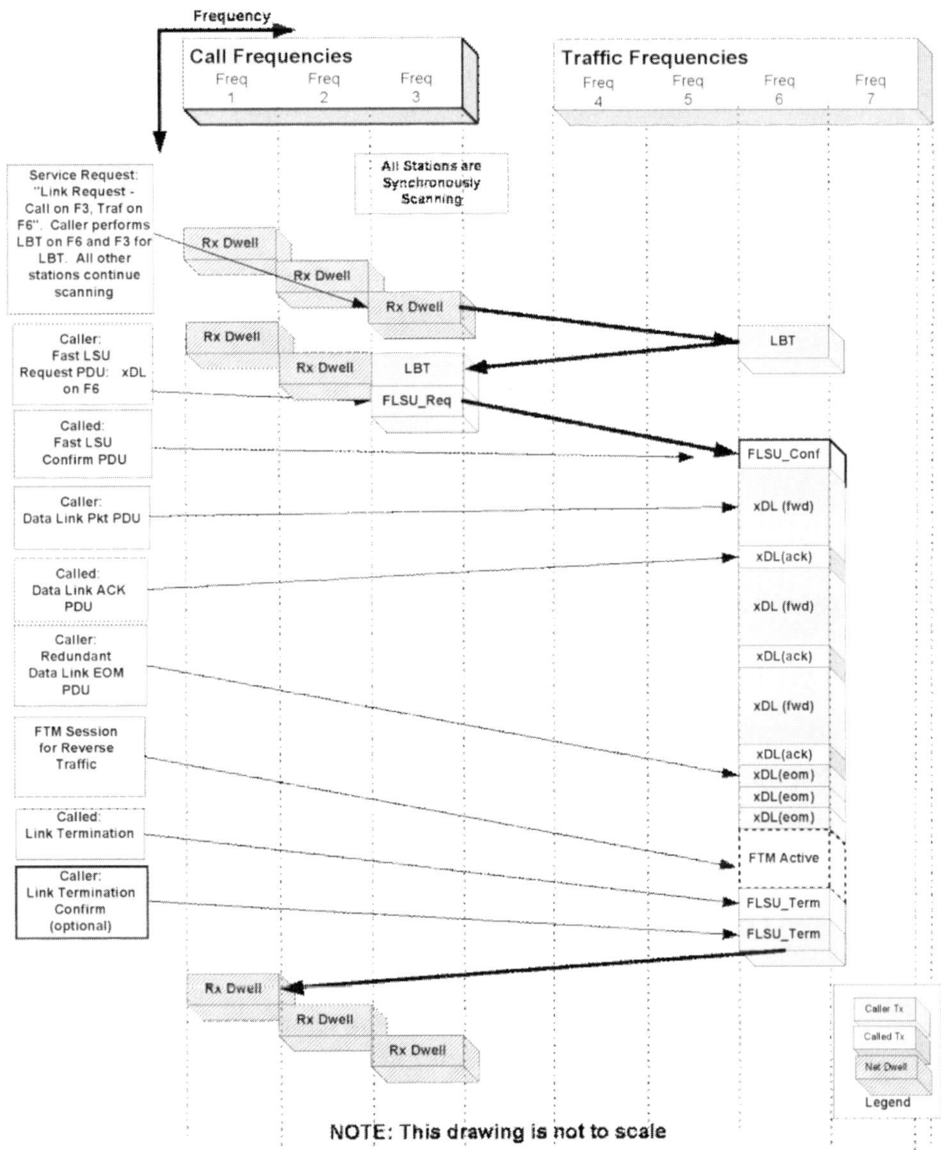

Figure 5.22 Separate calling/traffic channels in FLSU.

channel 3 and traffic channel 6. The station continues scanning until two dwells prior to dwelling on the desired calling channel (channel 3), at which time it switches to the desired traffic channel (channel 6), and performs an LBT for one dwell to assess occupancy. If the desired traffic channel is unoccupied, the station switches to the desired calling channel, performs LBT for one dwell, and then sends the link request PDU.

Both stations then switch to the traffic channel, and perform antenna tunes, if necessary. The called station issues a link confirm PDU on the traffic channel and the subsequent packet ARQ traffic proceeds as in the case for common calling and traffic channels.

Note that the link setup failure condition is identical to the common calling and traffic channel case. If the calling station does not receive the link confirm PDU on the traffic channel, it issues an ARQ EOM sequence (only if xDL mode was announced in the request), followed by a link termination PDU. The timing requirements are not changed for this scenario since they include time for switching frequencies and tuning whether the traffic channel is the same as the calling channel or not.

5.3.5 Robust Link Setup

This section describes the robust link setup (RLSU) protocol. RLSU is intended for use in large or heavily loaded 3G HF networks. It differs from FLSU in a few important ways, reflecting its orientation towards heightened efficiency for such applications:

- The RLSU PDU is much shorter than the FLSU PDU (26 bits versus 50), so a call occupies a calling channel for a shorter time.
- The RLSU dwell contains multiple time slots for calling, which allows for more effective collision avoidance than FLSU.
- Although both FLSU and RLSU can operate in either trunked or shared calling/traffic channel modes, FLSU normally shares channels while RLSU normally uses trunking.

Functionally, RLSU offers about the same capabilities as FLSU (PTP and PTM linking, time distribution, and so on), but its shorter PDU does not allow for traffic setup to occur concurrently with link setup. When RLSU sets up a link for analog voice traffic, there is no need for a traffic management handshake after link setup. However, for all digital modes (including digital voice), a traffic management (TM) handshake must follow a successful link setup handshake. This sequence is shown generically in Figures 5.23 and 5.24. Detailed examples are presented later.

An interesting difference between the link setup procedures of FLSU and RLSU lies in which station selects the channel to be used for traffic. In FLSU, the traffic channel is designated in the Request PDU, sent by the caller. In RLSU, the caller announces the type of traffic it wishes to send, and the *called* station selects a channel that it believes will support that type of traffic.

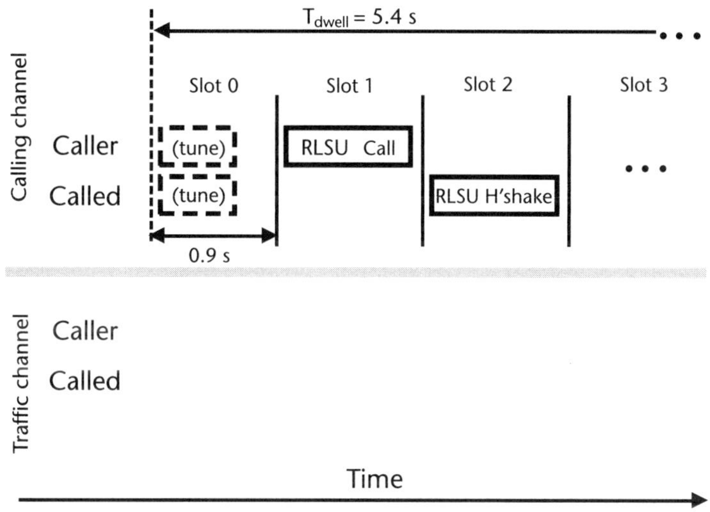

Figure 5.23 RLSU handshake.

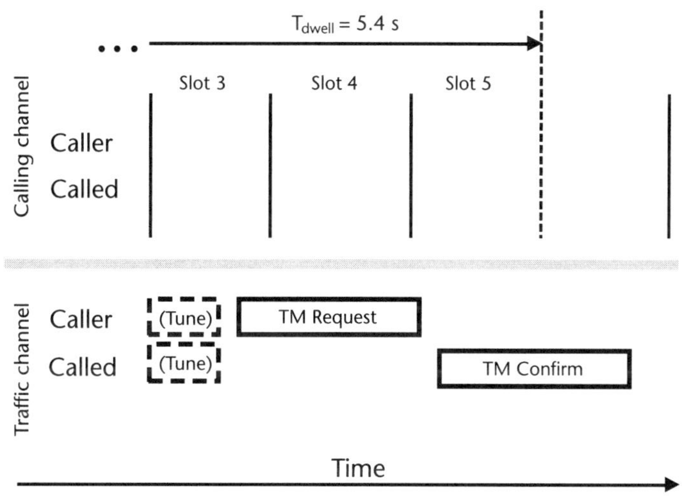

Figure 5.24 Traffic management handshake following RLSU handshake.

5.3.5.1 RLSU Protocol Data Units

The robust-mode link setup protocol data units (RLSU PDUs) are shown in Figure 5.25. In each point/multipoint address, the four least-significant bits are termed the "group #." The six most significant bits form a "member #" field. These PDUs are sent using BW0.

RLSU_Call PDU

The RLSU_Call PDU conveys necessary information to the called station(s) so that those station(s) will know whether or not to respond and what quality of traffic

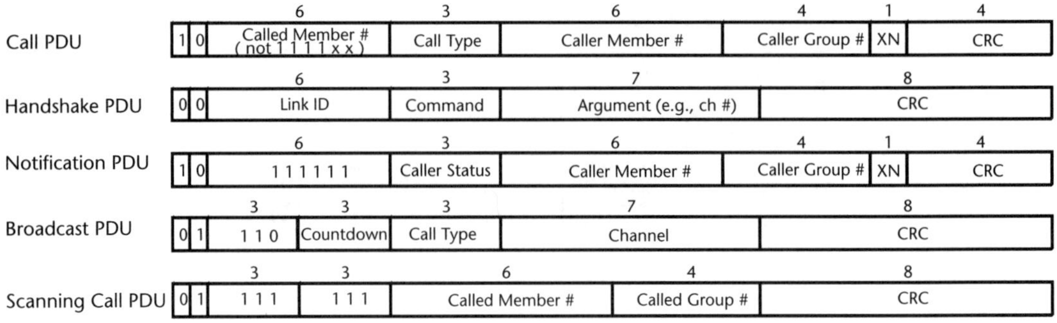

Figure 5.25 RLSU PDUs.

channel will be needed. The full address of the *caller* is included in the PDU, but only the member number field of the called station is needed. This is because the group number portion of the called station address is used to compute the calling channel that carries the call, and is therefore implicit. The XN (cross net) bit in the RLSU_Call PDU is set to '0' when the network numbers of the calling and called stations are identical. Otherwise the XN bit is set to '1'.

The call type field in the RLSU_Call PDU is encoded as specified in Table 5.7. The call types are described below:

- The *packet data* call type is used only when the HDL or LDL data link protocol will be used to deliver a message after link establishment.
- The *HF modem circuit* call type is used when an HF data continuous waveform (i.e., a waveform other than BW1–BW4) will be used to convey traffic after link establishment.
- The *voice circuit* call type requests a link SNR suitable for analog voice operation (for example, 10 dB or better).
- The *high-quality circuit* call type requests a substantially better channel than an analog voice circuit (for example, 20 dB or better), usually for carrying large amounts of data via a high-speed HF data waveform.

Table 5.7 RLSU Call Type Field Encodings

Code	*Call Type*	*Second PDU From*	*FromCall Type*
0	Packet data	Responder	Packet data
1	HF modem circuit	Responder	HF modem circuit
2	Analog voice circuit	Responder	Analog voice circuit
3	High-quality circuit	Responder	High-quality circuit
4	Unicast ARQ packet	Caller	Unicast ARQ packet
5	Unicast circuit	Caller	Unicast circuit
6	Multicast circuit	Caller	Multicast circuit
7	Control	Caller	Control

- The *unicast* and *multicast* call types are used when the calling station will specify the traffic channel used for a link, and are useful for called stations maintaining radio silence.
- The *control* call type is used to announce link release, to initiate a sync check handshake, and similar functions, rather than to establish a link.

RLSU_Handshake PDU

The RLSU_Handshake PDU is the second PDU sent in an RLSU handshake.

The *link ID* is a hash of the caller and responder addresses, computed as follows from the 10-bit point/multipoint addresses of the caller (node sending the RLSU_Call PDU) and the responder (called station or multicast address in RLSU_Call PDU):

temp1 = <caller address> * 0x13C6EF

temp2 = <responder address> * 0x13C6EF

LinkID = ((temp1 >> 4) + (temp2 >> 15)) & 0x3f

where '*' indicates 32-bit *unsigned* multiplication, '>> n' indicates right shift by n bits, and '&' indicates bitwise AND. Example LinkID computations are shown in Table 5.8.

The Command field is encoded as shown in Table 5.9.

The "argument" field contains a channel number, a reason code, or 7 bits of data, as indicated in Table 5.9. Reasons are encoded as 7-bit integers with values selected from Table 5.10.

RLSU_Notification PDU

The RLSU_Notification PDU is used to broadcast the status of the sending station: time server, leaving the network, commencing EMCON (radio silence), or nominal. Sending a notification of nominal status also serves as the 3G version of a sound. Notifications are always sent to the network to which the sending station belongs, so the XN bit in the RLSU_ Notification PDU should always be '0'.

Table 5.8 LinkID Computations

Caller	*Responder*	*temp1*	*temp2*	*LinkID*	*LinkID*
1	2	0013c6ef	00278dde	3D	61
1	3	0013c6ef	003b54cd	24	36
2	1	00278dde	0013c6ef	4	4
3	1	003b54cd	0013c6ef	33	51
(decimal)	(decimal)	(hexadecimal)	(hexadecimal)	(hexadecimal)	(decimal)

Table 5.9 Command Field Encodings

Code	*Command*	*Description*	*Argument*
0	Continue Handshake	The handshake will continue for at least another two-way handshake (on the next assigned called station dwell frequency when operating in synchronous mode).	Reason
1	Commence Traffic Setup	This is the final command sent on a calling channel. The argument is the channel number on which the responding station will (or should) listen for traffic setup. Following this command, all stations proceed to that traffic channel.	Channel
2	Voice Traffic	This command directs called station(s) to tune to a traffic channel and commence voice traffic. The argument is the channel number. Following this command, the calling station will be first to speak (PTP1 and PTM only).	Channel
3	Link Release	This command informs all listening stations that the specified traffic channel is no longer in use by the sending station.	Channel
4	Sync Check	This command directs the called station to measure and report synchronization offset back to the calling station. Used in Synchronization Management Protocol	Quality | Slot
5	Reserved		
6	Reserved		
7	Abort Handshake	This command immediately terminates the handshake and needs no response.	Reason

Table 5.10 RLSU Reason Field Encodings

Code	*Reason*	*Remarks*
0	NO_RESPONSE	Required resources not responding
1	REJECTED	Station is unwilling to link
2	NO_TRAF_CHAN	All traffic channels are in use
3	LOW_QUALITY	Available traffic channels are of insufficient quality to support requested traffic
others	(Reserved)	

RLSU_Broadcast PDU

The RLSU_Broadcast PDU may be used to establish broadcast links. The call type field describes the traffic to be sent, using the encodings in Table 5.7. The channel field contains the number of the channel on which the calling station will broadcast. The countdown field indicates the remaining time (in dwells) until the broadcast will begin.

Scanning Call PDU

The RLSU_Scanning_Call PDU is used for asynchronous link setup. The complete address of the called station is contained in this PDU. Scanning call PDUs are sent multiple times to capture an asynchronously scanning receiver, immediately followed by a single RLSU_Call PDU.

CRC Computation for RLSU PDUs

Each RLSU PDU contains either a 4-bit or an 8-bit CRC computed as follows:

- The 4-bit CRC uses the polynomial $x^4 + x^3 + x + 1$.
- The 8-bit CRC uses the polynomial $x^8 + x^7 + x^4 + x^3 + x + 1$.

5.3.5.2 *RLSU Protocol Operation*

As in any ALE network, idle stations in a 3G network using RLSU scan calling channels listening for calls (or sounds).

Synchronous Dwell Structure

Each synchronous dwell period in RLSU consists of six 900-ms slots, as shown in Figure 5.26. Slot 0 is reserved for tuning and for checking for occupancy of a traffic channel (selected in rotation as described later). Following the traffic channel occupancy check, the remainder of the dwell consists of 5 calling slots on the calling channel of 900 ms each for calling and notifications. Calls may be initiated during any of the first four calling slots during the dwell.

Calling Slot Selection

The calling station selects a calling slot for each dwell, using a random or pseudorandom process that is skewed according to the priority of the traffic to be sent. Earlier slots are less likely to be preempted by other calls than later slots, so higher-priority calls should be more likely to select earlier slots. A suggested approach to

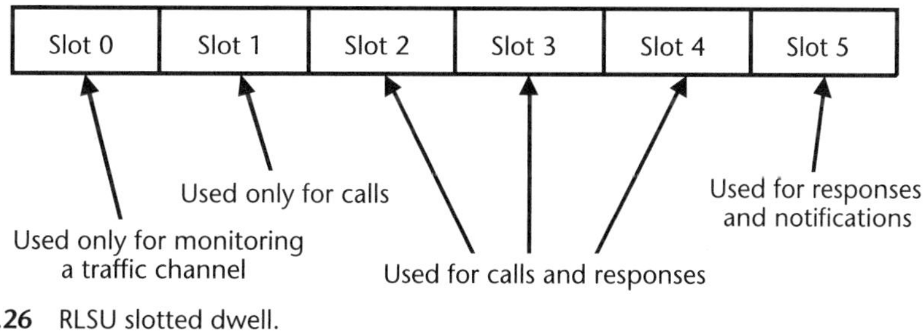

Figure 5.26 RLSU slotted dwell.

such prioritization uses the probabilities in Table 5.11. Note that calls are never sent in Slot 5 of a dwell, but a response to an earlier call may be sent in that dwell.

Synchronous Calling Protocols

A station begins calling upon request by its user, and continues calling in each subsequent dwell until either the logical link is established or the call is aborted by the user, the other station, or due to failure to link after a predetermined number of calls. RLSU can establish a link in one of three ways:

- *PTP RLSU with ACK (PTPA)*: the called station is expected to respond to the call and either accept the logical link and designate a traffic channel, reject the logical link, or accept the link but defer selection of a traffic channel.
- *PTP one-way RLSU (PTP1)*: the calling station will designate the traffic channel; no RLSU response is expected or allowed from the called station. This case is sometimes called *unicast calling*.
- *PTM one-way RLSU*: the calling station will designate the traffic channel; no RLSU response is expected or allowed from any of the called stations. This case comprises multicast (PTM1) and broadcast calling (BCST). PTM1 with acknowledgment uses the TM protocol roll-call after RLSU.

Listen Before Transmit

The avoidance of collisions and interference is a key requirement of 3G ALE. Stations employing RLSU must listen on the calling channel during the slots that precede the one selected for sending a Call. If signalling is detected during the listen before transmit (LBT) period, the station must not send a PDU in the following slot unless the following optional determination is made: the detected signalling is an RLSU_Handshake PDU with a correct CRC, and the command field indicates that traffic will not begin on the channel.

Robust PTP and PTM Calling

In both cases of PTP calling (PTPA and PTP1), as well as for PTM calling, each time the called station (or multicast group) begins a new dwell time, the calling station will select a slot for sending a call PDU, monitor the calling channel in the preceding

Table 5.11 RLSU Slot Selection Probability

	Probability of Calling in Slot			
Traffic Priority	1	2	3	4
Highest	50%	30%	15%	5%
High	30%	45%	15%	10%
Routine	10%	15%	45%	30%
Low	5%	15%	30%	50%

slot (if any), and send the call if LBT indicates that the slot will be free. However, if the calling station receives a call addressed to itself, the station should abandon its own call and respond to the incoming call.

If LBT indicates that the channel may be busy, the station will not send a call during that dwell on that channel and the station should instead listen for calls addressed to itself during the remainder of that dwell.

Robust PTPA Responses

In PTPA RLSU, the responsibility for selecting a traffic channel rests with the called station, because that station can measure the current state of propagation on the calling channel and thus estimate the quality of associated traffic channel or channels (i.e., traffic channels in the same frequency band as the calling channel). The called station identifies a suitable traffic channel for a logical link by combining the signal quality of current (and recent) measurements of the current (and recent) physical links to the calling station with occupancy measurements of associated traffic channel(s). It then compares the resulting estimated traffic channel quality with the minimum channel quality required for the grade of service indicated in the call. The called station may use any suitable algorithm for selecting a traffic channel from those channels that meet the required minimum channel quality. This channel need not be in the vicinity of the current calling channel.

Upon receipt of a PTPA call addressed to it, a station that is searching for calls will send one of the following responses:

- *Commence Traffic*: the response identifies the traffic channel on which the calling station is expected to initiate the traffic setup (TM) protocol.
- *Voice Traffic*: the response identifies the traffic channel on which the calling station is expected to initiate analog voice traffic.
- *Continue Handshake*: indicates that the responding station is willing to establish the requested logical link, but wishes to continue handshaking on calling channels to collect more propagation measurements. The called station makes no state change after sending this response, but continues to search for calls as before.
- *Abort Call*: indicates that the responding station is not willing to establish the requested logical link and contains a code that explains the reason.

The response is sent on the channel that carried the call in the calling slot that immediately follows the call.

Robust PTP1 and PTM1 Responses

In PTP1 and PTM1 RLSU, the responsibility for selecting a traffic channel rests with the calling station, which must estimate traffic channel quality using a combination of predictions and measurements (when available) for the physical links to the intended recipient(s) of the traffic. The calling station sends the response (RLSU_Handshake PDU) on the channel that carried the call in the slot that immediately follows the call. Although any of the four responses listed above for the PTPA case are allowed, the response will normally be *commence traffic* or *voice traffic*, and will identify the traffic channel on which the calling station will initiate the TM protocol or an analog voice transmission.

Conclusion of Robust LSU

After a call and a commence traffic response have been sent, the calling station and all called stations that have sent or received the call and response will tune to the indicated traffic channel and begin the traffic management protocol. Note that when a link is established for voice traffic (Call Type in RLSU_Call PDU = Analog Voice and Command field in RLSU_Handshake PDU = Voice Traffic), the TM protocol enters the appropriate linked state without a TM handshake.

Called stations set a timeout that will return them to searching for calls if traffic does not begin in a timely manner. If the calling station did not receive the response in a PTPA call, it will, of course, not begin traffic setup. It will instead proceed to the next calling channel to continue calling.

Robust LSU Synchronous Mode Link Release

At the conclusion of an individual PTPA, PTP1, or unicast PTM link, the caller may optionally send a link release. A link release comprises an RLSU_Call PDU containing the original called station address, with a call type of control, followed by an RLSU_Handshake PDU that identifies the traffic channel and contains a link release command.

The link release is sent on the *calling* channel on which the handshake that set up the link occurred. The calling station should attempt to send the link release during the first dwell after the link is terminated, during which the called dwell group is listening on that calling channel. The calling station need not attempt to send a link release later if calling channel occupancy during that dwell prevents transmission of the link release. The slots used for a link release are selected randomly using the probability distribution for lowest priority calls.

Note that scanning stations that track links established and released must attempt to interpret all PDUs. This results in additional computational burden on scanning stations.

Robust LSU Synchronous Mode Broadcast Calling

An RLSU_Broadcast PDU directs every station that receives it to a particular traffic channel, where another protocol (possibly voice) will be used. A means is typically provided for operators to disable execution of the broadcast protocol.

The call type field in the RLSU_Broadcast PDU is encoded as in the RLSU_Call PDU, except that only the circuit call types may be used.

The countdown field indicates of the number of dwells that will occur between the end of the current dwell and the start of the broadcast. A countdown value of 0 indicates that the broadcast will begin in Slot 1 of the following dwell. Other countdown field values n > 0 indicate that the broadcast will begin no later than n dwell times in the future. The countdown field is be decremented for each new dwell during a broadcast call.

The channel field indicates the channel that will carry the broadcast.

A station may send an RLSU_Broadcast PDU in every slot in a dwell (except for Slot 0). It may also change channels every slot to reach a new dwell group. While sending RLSU_Broadcast PDUs (only), the calling station does not need to check occupancy on the new calling channel before transmitting on that channel.

At the beginning of Slot 1 in the dwell after the caller sent RLSU_Broadcast PDU(s) with the countdown field set to 0, the caller commences TM (or voice) on the indicated channel.

Stations that receive an RLSU_Broadcast PDU and tune to the indicated traffic channel return to scan if the TM_Request (or voice transmission) does not begin within the traffic wait timeout period after the announced starting time of the broadcast.

RLSU Examples

Figures 5.27 through 5.30 describe, by means of specific example scenarios, a subset of the capabilities provided by the Robust LSU protocol. The following scenarios are described:

- Synchronous two-way RLSU, point-to-point packet service;
- Synchronous two-way RLSU, point-to-point circuit mode service for analog voice;
- Synchronous broadcast RLSU for an analog voice broadcast;
- Asynchronous two-way RLSU, point-to-point circuit mode service for analog voice.

All of the scenarios show a two-dimensional (time and frequency) view, with four search frequencies and four traffic frequencies listed on the horizontal axis, and time on the vertical axis (time progresses from top to bottom). The search frequencies and traffic frequencies are distinct (trunked operation). In the channel database, the calling channels would be numbered 0 through 3, and the traffic channels would be numbered 4 through 7.

A legend depicts how stations are identified : light gray depicts the caller activities, white depicts called station activities, and dark gray (cross-hatch) depicts the activities of all of the net members.

In these examples, all of the stations scan the same channels at the same time, so all are in the same dwell group. In large or busy networks, stations would be assigned to multiple dwell groups in order to reduce calling channel congestion.

The partition of the available frequencies into four calling channels and four traffic channels indicates that the network manager expects traffic to consist of frequent, short messages. When each call would result in extended use of a traffic channel, the partition would be skewed to increase the number of traffic channels relative to the number of calling channels.

Synchronous Two-Way RLSU, Point-to-Point Packet Service

Beginning at the top left corner, Figure 5.27 shows that all stations in the net synchronously scan the assigned frequencies. The dwell period is 5.4 seconds per frequency. While scanning, all the stations are required to perform the listen before

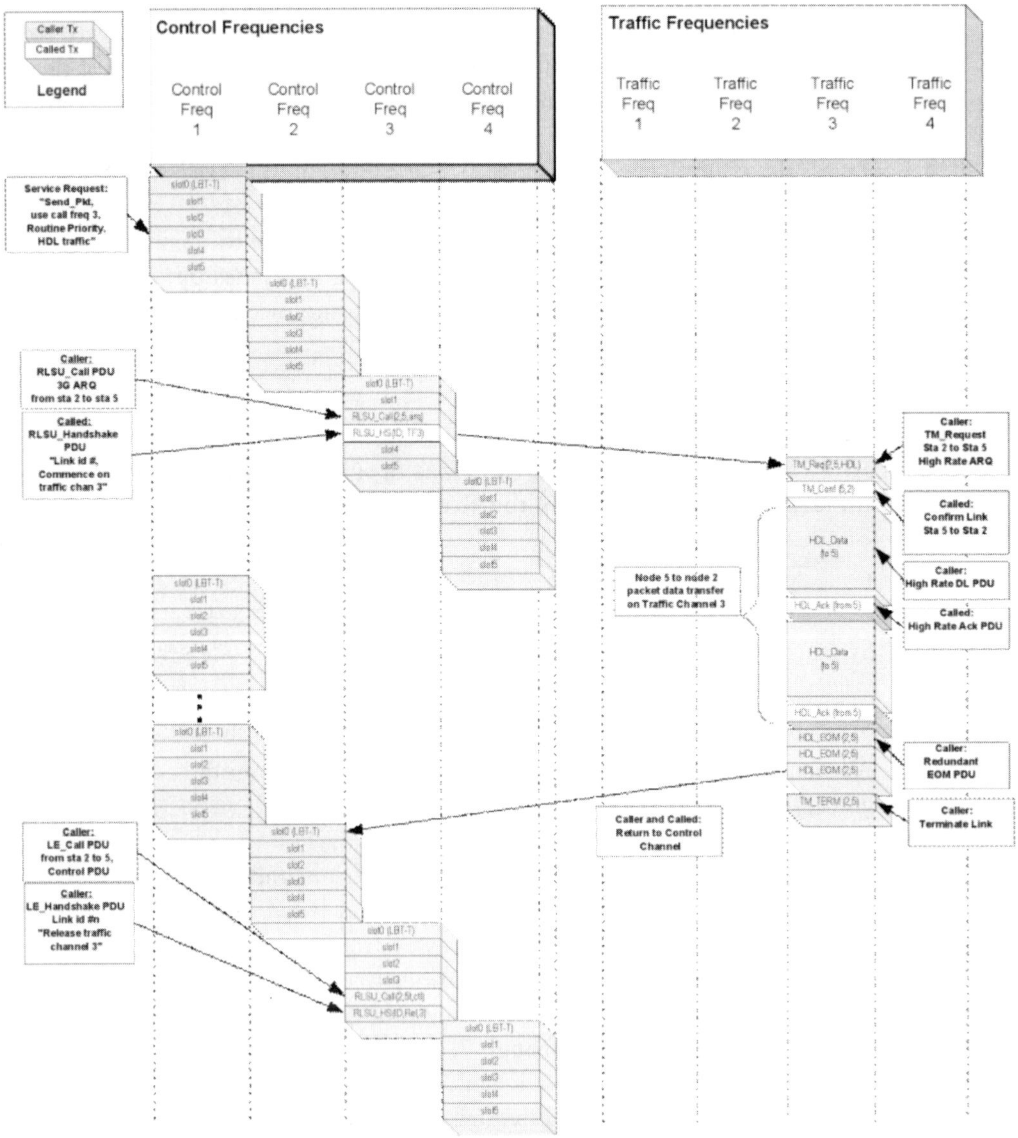

Figure 5.27 Synchronous two-way RLSU, point-to-point packet service.

transmit (LBT) algorithm during the first portion of each dwell (Slot 0) as a means of establishing a frequency occupancy perspective for each of the frequencies in the scan list.

During the dwell on frequency 1, station 2 is directed to establish a point-to-point link with station 5, calling on frequency 3 (F3), using the xDL (generic reference to either HDL or LDL) ARQ protocol for reliable packet transfer.

The caller continues scanning until the *called* station is dwelling on the desired calling frequency. During this period, the caller is still available to respond to higher and equal priority received calls. If such a preemption occurs, then the original intended call is deferred. Otherwise, at the beginning of the dwell on F3,

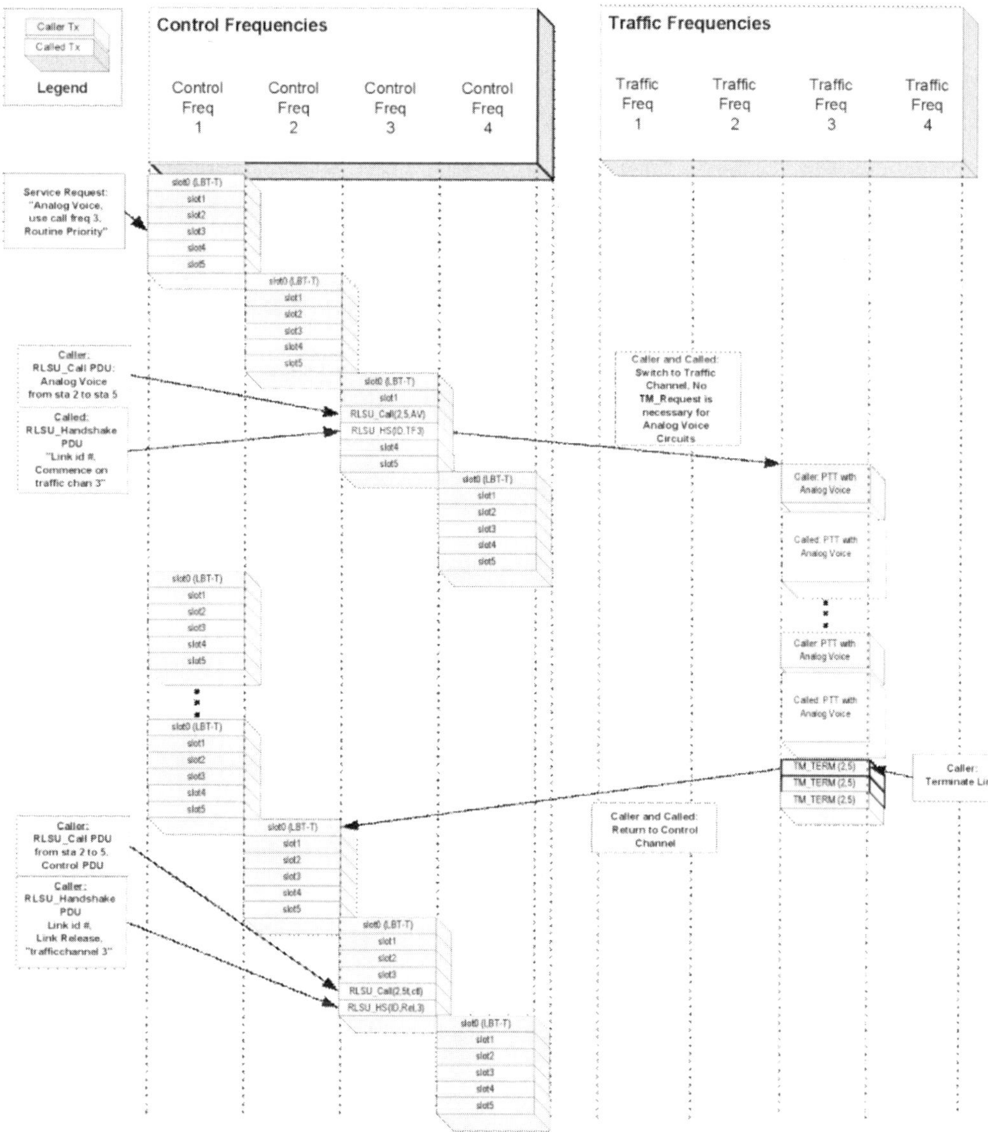

Figure 5.28 Synchronous two-way RLSU, point-to-point voice circuit service.

the caller station randomly selects a slot for its call (considering the priority of its call), executes a listen before transmit (LBT) process to assure that the channel is unoccupied during the slot preceding its selected slot, and sends a PTPA call in its selected slot (specifying a traffic type of packet or 3G ARQ).

The remaining stations are all dwelling on F3 and will detect the call if a physical link exists from station 2. In this example, station 5 receives the call and responds with an RLSU_Handshake PDU that accepts the call and designates traffic frequency 3 for the packet transfer. Again, other stations may receive the handshake PDU; stations that receive both the call and the response may note that the two stations and traffic channel 3 are now in use, and defer calls to those stations

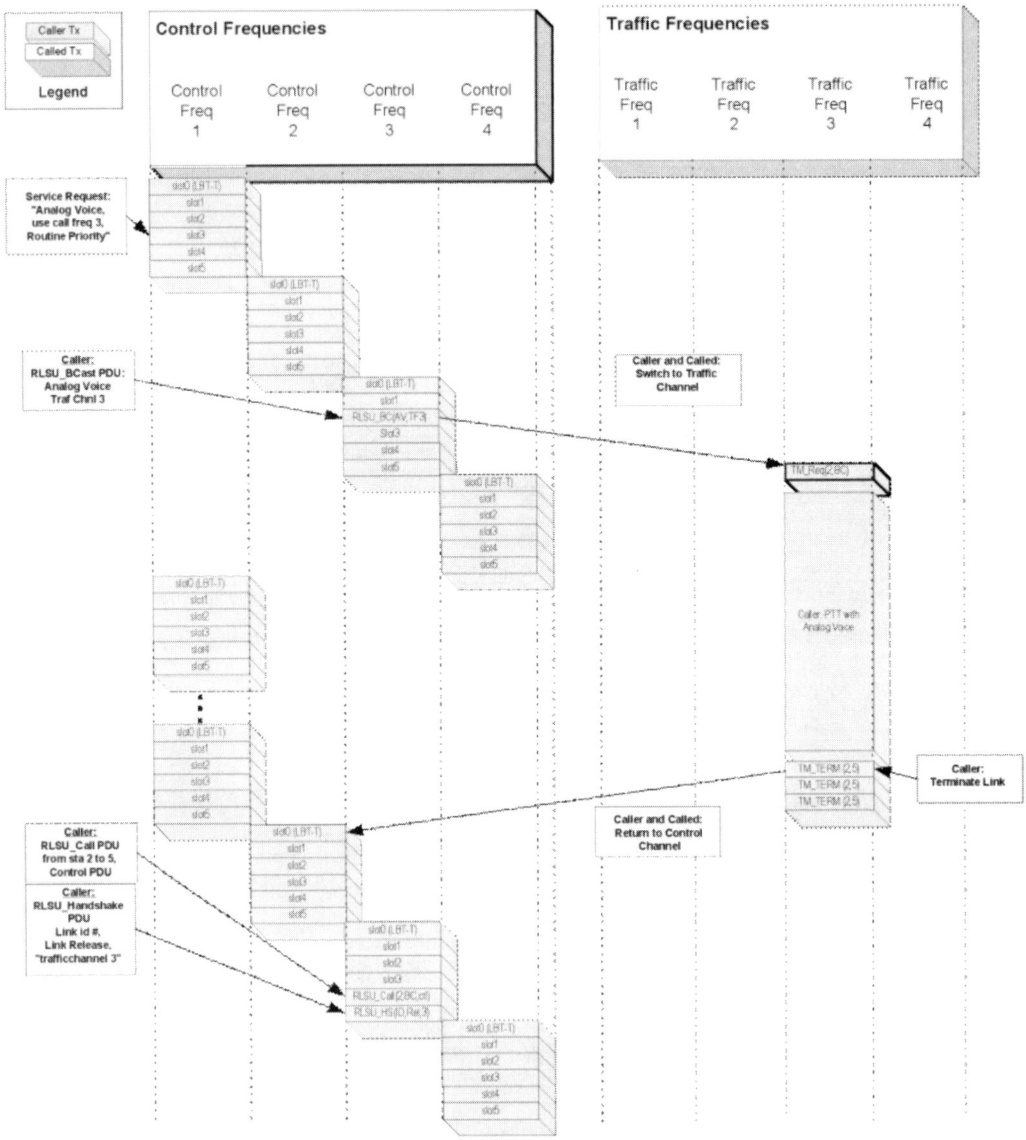

Figure 5.29 Synchronous broadcast RLSU, analog voice broadcast service.

and use of that traffic channel until a positive indication is received that that link has ended (or a timeout expires).

Stations 2 and 5 tune to traffic frequency 3, and the caller station (station 2) issues a TM_Request PDU that conveys the caller station address, called station address, priority, and the desired traffic service (HDL ARQ mode with a specific frame length). Station 5 returns a TM_Confirm PDU that completes the traffic setup phase, and the packet transfer commences using HDL ARQ.

Both the caller and called stations alternate sending HDL PDUs, with the caller sending data using the HDL data send PDU, and the called responding with the HDL ACK/NAK PDU. This process continues until all data has been transferred er-

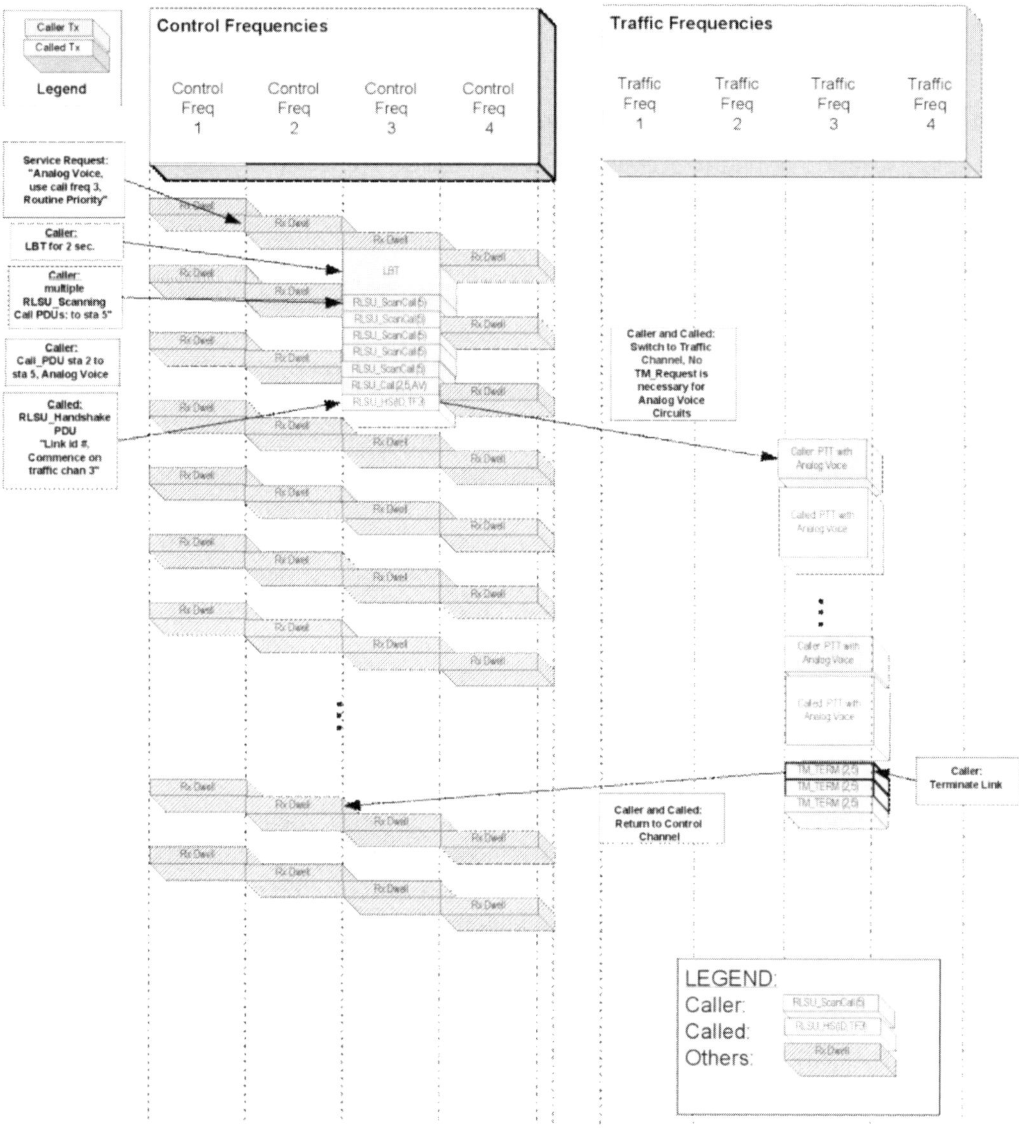

Figure 5.30 Asynchronous two-way RLSU, point-to-point voice circuit service.

ror-free, as indicated by the caller sending redundant HDL end of message (EOM) PDU's.

In a packet connection, the link is typically terminated immediately after the packet is successfully transferred, after allowing time for the called station to initiate sending a packet in the reverse direction. The caller initiates link termination by transmitting a TM_Terminate PDU. (The called station may optionally confirm link termination by echoing the TM_Terminate PDU.) After terminating the link, both the caller and called rejoin the other net members in synchronous scanning. During the first dwell of the called station on the calling channel that carried the successful LSU handshake, the caller attempts to send a *link release sequence* (control type RLSU_Call PDU followed by a link release command in an RLSU_Handshake PDU) to notify other stations that the link has been terminated.

Synchronous Two-way RLSU, Point-to-Point Voice Circuit Service

The scenario in Figure 5.28 is identical to the above scenario, except that the traffic service is *voice circuit mode*. Note that no TM handshake is performed on the traffic channel; instead, a voice conversation commences immediately after the stations tune to the traffic channel.

The same procedure is followed to terminate the link and to announce its release to other stations.

Synchronous Broadcast RLSU, Analog Voice Broadcast Service

The scenario in Figure 5.29 is identical to the above scenario except that the link is set up in *broadcast mode*. A single RLSU_Broadcast PDU is sent on calling channel 3, which specifies that a voice broadcast will begin immediately on traffic channel 3. All stations that receive the call immediately tune to traffic channel 3 to receive the broadcast from station 2. As usual, the caller terminates the link and announces its release to other stations.

Asynchronous Two-Way RLSU in Asynchronous Network, Point-to-Point Voice Circuit Service

The scenario depicted in Figure 5.30 is different from the preceding examples. The network is shown operating in asynchronous mode, wherein stations scan asynchronously the assigned calling channels at a rate of 787.5 ms per dwell (approximately 1.27 channels per second). This mode of operation may be selected when synchronous operation is impractical.

When the RLSU process at station 2 receives the request to call station 5 on calling channel 3, it immediately begins the link setup procedure since it has no knowledge of the channel currently monitored by station 5. The asynchronous call begins with the LBT (for 2 seconds by default), followed by the transmission of about 1.56C RLSU_Scanning_Call PDUs on the requested calling channel, where C is the number of channels in the scan list, followed by an RLSU_Call PDU. Transmitting 1.56C RLSU_Scanning_Call PDUs guarantees that all other scanning stations will scan the calling channel during the async call. Because the address of the called station(s) is contained in the RLSU_Scanning_Call PDU, all stations that are not included in the call are free to continue scanning. Called station(s) that receive one of the scanning call PDUs stop scanning and wait for the normal RLSU_Call PDU, which is sent immediately after the final Async_FLSU_Request PDU. After receiving a valid RLSU_Call PDU, the RLSU protocol proceeds as for the synchronous case, except that the response is sent a fixed time after receipt of the call. The link release protocol is optional in asynchronous mode and is not used in the scenario in Figure 5.30.

5.3.6 3G ALE Performance

The 3G ALE protocols have been thoroughly evaluated in the laboratory, in simulation, and in the field. In this section, we present many of the results of that performance evaluation.

5.3.6.1 Laboratory Measurements of BW0 and BW5 Linking Performance

The accepted method of testing ALE and other HF systems uses baseband channel simulators that implement the Watterson model of the HF channel. In this model, a signal is simulated as travelling through the HF channel via two independently fading paths with a fixed time offset between them. The two paths have equal average path loss. Noise is modeled as additive white Gaussian noise (AWGN). The settings on such a channel simulator are the SNR of the arriving signal, the time spread between the two paths, and the two-sigma fading bandwidth of the paths. The usual channel conditions used in testing HF modems, ALE systems, and so on are as follows:

- A *Gaussian channel* has only a single, non-fading signal path. This is intended to represent surface-wave propagation.
- A *good channel* has 0.5 ms path spread, and a fading bandwidth of 0.1 Hz. This models a slowly fading skywave channel.
- A *poor channel* has 2.0 ms path spread and a fading bandwidth of 1 Hz. Despite the discouraging name, this represents fairly typical mid-latitude skywave channel conditions. (Such channel conditions are termed *mid-latitude disturbed* in ITU-R Recommendation F.1487 [5]).

Table 5.12 lists the required probability of successfully linking as a function of the type of channel and the SNR on that channel. Both FLSU and RLSU systems meet these requirements.

Figure 5.31 shows the required linking performance of 2G and 3G ALE systems in these three standard channels. The 3G system achieves 8 to 10 dB better performance than the 2G system in the standard channels, and is also able to link under narrowband interference that would prevent a 2G system from linking.

What are the implications of the improved SNR performance of 3G systems? In strategic and long-haul HF applications, we generally find high-power transmitters

Table 5.12 LSU Linking Probability Requirements

Probability of Linking Success (%)	*Gaussian Channel*	*Channel Good*	*Channel Poor*
25	−10	−8	−6
50	−9	−6	−3
85	−8	−3	0
95	−7	1	3

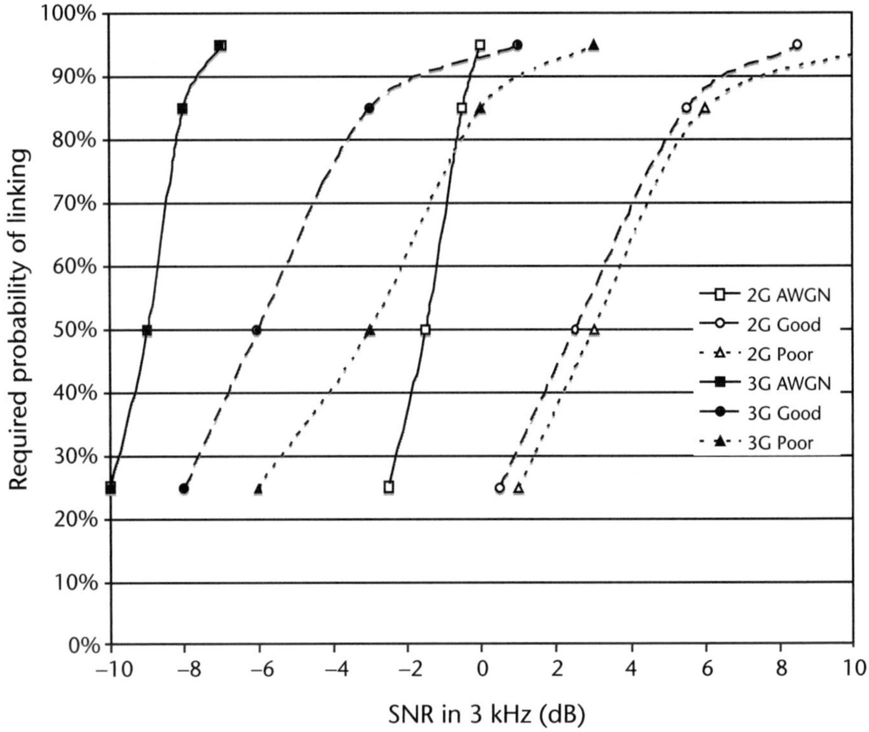

Figure 5.31 Comparison of linking performance requirements.

connected to external electrical power. Coverage is a primary concern, so the transmitters are often operated at the maximum power permitted by ITU regulations. In these applications, the extra SNR performance extends the available coverage area or reachback distance. For example, a 3G base station can reach an aircraft flying at greater distances than a 2G base station can.

For tactical applications, we find qualitatively different benefits of a better SNR performance. Tactical radios are often powered by batteries carried with the radio. In such applications, the ability to communicate at lower SNR means that transmit power can be reduced, which prolongs the life of batteries (a clear benefit if you can carry fewer batteries in your pack).

5.3.6.2 Simulation Study of 3G ALE Performance

We have gauged the effectiveness of 3G ALE in HF network applications using simulation studies using the NetSim simulator [6]. This family of simulators is considered fairly reliable, as it has been independently validated (in its 2G version) by the Joint Interoperability Test Command [7]. We here summarize results from two scenarios previously reported at MILCOM '99 [8]: (a) a strategic-scale air-to-ground scenario, in which a global network of HF base stations provides voice communications to aircraft flying typical transport missions, and (b) a tactical scenario, in which network radios exchange short data messages.

Air-to-Ground Scenario

The air-to-ground scenario models a large-scale HF network used for voice communications. A fleet of aircraft flies among bases on the east coast of the United States, Europe, North Africa, and South Asia, as shown in Figure 5.32. The original Walnut Street model (discussed in Chapter 2, Section 2.4.3) is used for skywave propagation, during the month of June, with a sunspot number of 100. For comparison, identical scenarios were simulated using 2G and 3G radios. The spectrum allocated for this relatively large network included 18 frequencies spread over most of the HF band. Various combinations of calling and traffic channels were simulated with the 3G radios, and the best combination is reported in the results below:

- All antennas were omnidirectional with 0 dBi gain.
- Each of 14 ground stations spread around the globe had two identical radios, either 2G or 3G RLSU, each with a 4-kW power output.
- Each ground station sent a sound on each calling channel every 45 minutes.
- Each aircraft carried a single radio, either 2G or 3G RLSU, with a 400W power output.
- The aircraft did not sound.
- The aircraft schedule was selected to be typical of U.S. Air Force Air Mobility Command operations in the 1990s (when the 3G technology was in development).

Figure 5.32 Air-to-ground scenario. (After: [8].)

- Each aircraft placed (on average) one 5-minute voice call per hour while in the air. (Intervals and durations were exponentially distributed.) A ground station was selected for each call based on how the sounds they transmitted were received by the aircraft.

The adaptive calling by aircraft (covered in the last bullet) is an interesting aspect of this scenario, and was the cause for sounding by the ground stations (both 2G and 3G).

For this strategic voice application, the metric of interest to users (pilots) is the time required to set up the link. The cumulative percentage of all calls completed within 10, 20, 30, 60, and 90 s is plotted in Figure 5.33 for 2G and 3G ALE networks under identical loading. In the 3G network, the 18 frequencies were optimally partitioned: 5 frequencies were allocated for calling channels, and the other 13 were allocated for traffic. The 2G network used the same 18 frequencies for both ALE and traffic.

Just over 18 seconds was the minimum time required for the 2G ALE network to set up a link, so it completed no links in less than 10 seconds. However, the 3G ALE network completed roughly half of its links in less than 10 seconds, and in general, was significantly faster than 2G ALE in establishing links. This was due to both the shorter 3G ALE calling transmission and the more robust waveform that permitted the 3G system to make contact on low-SNR calling channels. When the call succeeded on a channel whose SNR was too poor for voice traffic, the 3G system could then redirect the link to a channel suitable for voice communication.

How well did having separate calling and traffic channels work in the 3G system? Utilization of the traffic channels during the busy hours averaged 28% to 49%, with some hourly channel utilizations of 83%. Utilization of the five calling channels, on the other hand, was very low, ranging from 1% (sounding only) to a maximum of 4%. Such low occupancy of these contact channels allowed most calling stations to place a call in every dwell until a link was established. In this

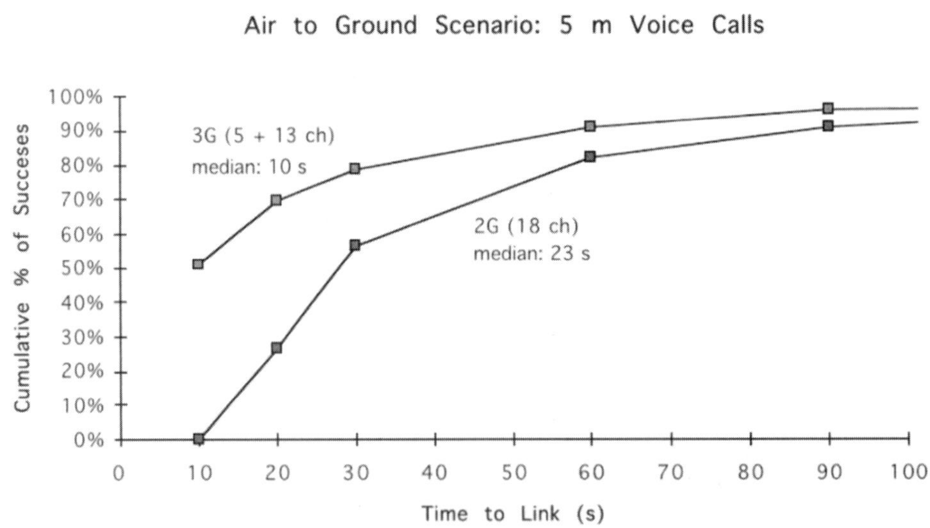

Figure 5.33 Air-to-ground linking time results. (© 1999 IEEE. Reprinted with permission from [8].)

simulation, the decoupling of traffic and calling channels was demonstrated to reduce the back pressure of heavy traffic on linking.

Tactical Mesh Network Scenario

For the tactical scenario, we investigated data message throughput in 3G mesh networks (i.e., peer-to-peer networks in which any station may call any other station). First, a 10-station network was simulated in which each station called the other nine with uniform probability. A larger 100-station network was used to investigate the scalability of the 3G technology. In this case, each station exchanged traffic only with its four nearest neighbors. Sounding was not used.

The total traffic volume presented to the network was varied over a range of 100 through 2500 messages per hour, with exponentially distributed interarrival times. No data link protocol was simulated; instead, the channel was occupied after linking for fixed periods of 5, 10, 15, or 30 seconds.

For comparison with previous studies, a fixed-SNR channel model was used: all links had path losses to produce 11 dB SNR (in the absence of collisions from other network members). Eighteen channels were allocated. Due to the comparable durations of linking and traffic, we selected an equal split of 9 calling and 9 traffic channels.

Figure 5.34 shows the message throughput achieved versus the offered message load in the 10-station mesh network. We see the asymptotic saturation expected in a CSMA/CA network. Previous work (unpublished) showed that 2G ALE networks without listen-before-transmit (LBT) exhibited ALOHA-style breakdown in throughput under high loads, and that 10-station 2G ALE networks *with* LBT saturated at about 15 messages per hour for 5 second messages [9].

Digging into the detailed results, we find that the single ALE radio at each station is the bottleneck in the 10-station network. For example, at an offered load of 250 messages per hour (25 messages per hour per station), all of the stations were linked at least 60% of the time, and 7 of the 10 stations were linked for more than

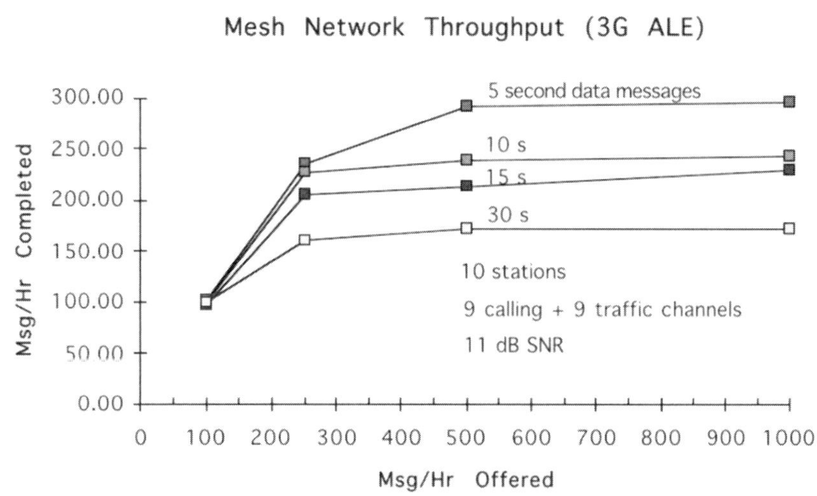

Figure 5.34 Message throughput in 10-station 3G mesh network. (© 1999 IEEE. Reprinted with permission from [8].)

85% of the time. Thus, many of the messages offered to the network could not be delivered because the destination station was not available to receive them. Nevertheless, the saturated 3G network sustained a message throughput that was more than an order of magnitude higher than for a comparable (10-station) 2G network.

Figure 5.35 illustrates the ability of 3G technology to scale up by an order of magnitude (100 stations). Even at an offered load 10 times greater than the saturation throughput of the 10-station network, the 100-station mesh network has not saturated. However, keeping the number of channels fixed apparently prevented the 100-station network from achieving 10 times the throughput of the smaller network. Calling channel utilization ranged from 24% to 27%, while traffic channel utilization ranged from 31% to 74%.

Station utilization in the 100-station network ranged from 1% to 14% at an offered load of 250 messages per hour (2.5 messages per hour per station), and was somewhat lower at 25 messages per hour per station than the corresponding figures for the 10-station network.

5.4 Traffic Management

Traffic management (TM) refers to the negotiation (or renegotiation) of how an established HF link will be used to deliver traffic, as well as explicitly terminating the traffic connection. The protocols used differ for FLSU and RLSU networks. The initial traffic management handshake in FLSU networks is included during the link setup handshake; subsequent negotiations use the FTM protocol. RLSU networks use a separate TM protocol, which is described in this section.

The transition to traffic following RLSU proceeds as follows, depending on the call type:

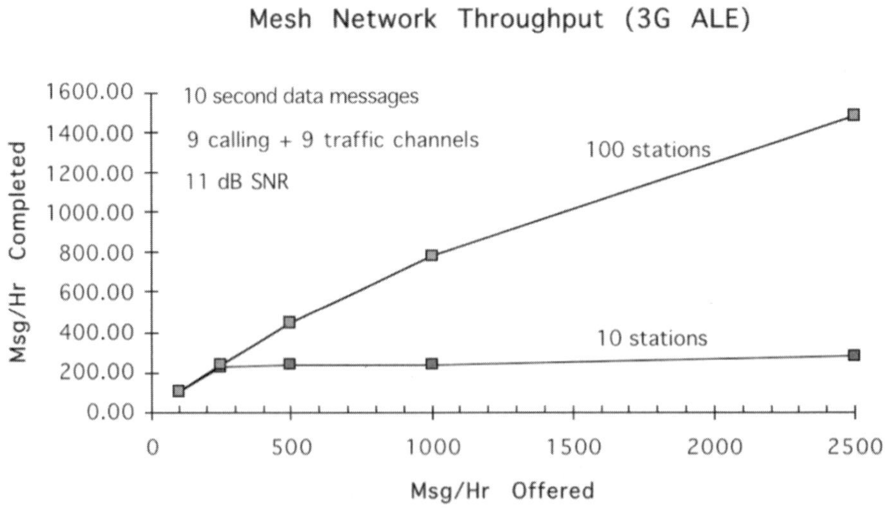

Figure 5.35 Message throughput comparison 10- versus 100-station 3G mesh networks. (© 1999 IEEE. Reprinted with permission from [8].)

- When an *analog voice call* succeeds, TM is not needed; voice traffic begins immediately.
- When a *packet data call* succeeds, the traffic setup (TM) protocol is engaged to set the parameters of the packet transfer.
- When a *circuit call* of any type succeeds, the circuit link control protocol governs transmissions on the link. Modems with autobaud capability will normally commence traffic without the necessity for traffic setup handshakes.
- The TM roll-call function is often useful for links established without acknowledgment from the called station(s).

Once a connection has been established, the stations participating in it determine:

- The identities of the stations intended to participate in the connection;
- The connection topology: point-to-point, multicast, or broadcast;
- The link mode: packet or circuit;
- The HF channel that will be used for signaling within the connection.

In addition, each participating station knows whether or not it initiated the connection (even though stations other than the initiator do not always know which station originated the connection, as in broadcast connections). The initiating station knows that it can transmit a TM PDU in the first transmit time-slot of the TM phase.

During the TM phase, the participating stations exchange TM PDUs in order to determine whether data or voice traffic will be carried, if the link is a circuit connection; which data link protocol(s), waveform(s), and/or baseband modulation formats will be used to deliver traffic on the connection; the priority of the traffic to be delivered; and the fine time synchronization required for the HDL and LDL protocols, on traffic links established for packet traffic.

If the traffic link is a *multicast circuit link* (has a multicast topology), the participating stations initially conduct a roll-call procedure to determine which of the stations in the multicast group received the RLSU signaling and are now present on the traffic frequency. A second roll-call can be conducted on the traffic link just before the traffic link is torn-down and the participating stations resume scanning. This allows a station sending information on a multicast circuit link to know whether the intended recipients of the information were on the traffic frequency to receive it. It also allows the station initiating the traffic link to drop the current link and attempt to reestablish it if desired stations are absent from the link.

When traffic exchanges have been completed on a traffic link, the TM protocol is used to coordinate the participating stations' departure from the traffic link.

5.4.1 TM PDUs

Traffic management PDUs are carried in BW1 bursts, and are formatted as shown in Figure 5.36. The fields are as follows:

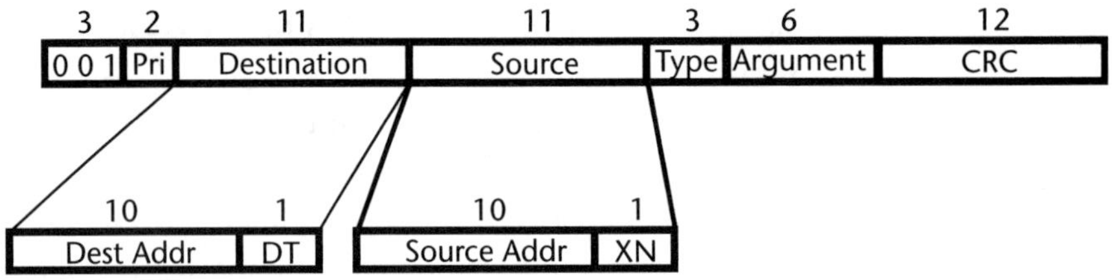

Figure 5.36 TM PDUs.

- *Priority* uses the usual 3G numbering scheme, with 0 as the highest priority.
- The *address fields* use 10-bit point/multipoint addresses. The destination type (DT) bit is 1 for a multipoint link, 0 for point-to-point. The XN bit is 1 if the traffic source network number is different from that of the destination, 0 if the same network.
- The *type field* specifies the role this PDU plays in the TM protocol: 0 for a TM request, 1 for a TM confirm, and 2 for a TM terminate.
- The *Argument field* in TM request and confirm PDUs carries the traffic type (see Table 5.5). In a TM terminate this field carries a reason code (Table 5.13).
- The *12-bit Cyclic Redundancy Check (CRC)* is computed over the preceding 36 bits of the PDU, using the polynomial $X^{12} + X^{11} + X^9 + X^8 + X^7 + X^6 + X^3 + X^2 + X^1 + 1$.

5.4.2 TM Protocol Operation

An example of the TM protocol in operation following RLSU is shown in Figure 5.24 in Section 5.3.5. After the initial TM handshake, the TM entity at each station continues to monitor the channel while the station is exchanging voice or data traffic. It must be ready to renegotiate traffic parameters or to drop the link upon receipt of a TM PDU. On traffic links established for packet traffic using the HDL or LDL protocol, the user process can terminate the data link transfer and use the next data link transmission time slot in either direction (i.e., the time slot for the xDL_DATA or the xDL_ACK PDU) to instead send a TM_TERM PDU. This means that while a data link transfer is in progress, each station must be simultaneously attempting to demodulate TM_TERM PDUs conveyed by the BW1 waveform as it is attempting to demodulate and receive data link signaling conveyed by BW2, BW3, or BW4. Similarly, on a circuit link, each station must attempt to detect and demodulate TM_TERM PDUs conveyed by the BW1 waveform at all times when the station is not transmitting.

In addition to selecting a traffic mode for a link, the TM protocol also establishes time synchronization for the HDL and LDL protocols. The TM phase of the point-to-point packet link is considered to begin at the end of the RLSU time-slot in which the LSU_COMMENCE PDU is transmitted. Since only a two-way TM handshake is performed, it is not possible for both stations to estimate the propagation delay between them. Instead, in each direction, the TM handshake signaling is used to establish the timing for all subsequent signaling in that direction. In the

Table 5.13 TM Terminate Reason Codes

Reason Code	*Description*
0 ("ABORT")	Immediately terminate the traffic link, with all participating stations leaving the traffic frequency or frequencies assigned to the link. Reason = ABORT indicates nothing about any measures that may be taken to resume any data transfer that may have been in progress.
1 ("RELINK")	Immediately terminate the traffic link, with all participating stations leaving the traffic frequency or frequencies assigned to the link. Reason = RELINK indicates that the user process will attempt to resume the data transfer, possibly on a different frequency or frequencies.
2 ("SIGN_OFF")	The station sending the TM_TERM PDU is departing the multicast circuit link, of which it is not the initiator. If two or more stations remain on the link, they may continue to exchange traffic.
3 ("UNLINK")	Sent by the initiator of a multicast circuit link, to cause the link to be torn-down after a final roll call is performed.
4 ("UNL_ACK")	Sent by a participant in a multicast circuit link in response to a TM_TERM PDU with Reason = UNLINK, to indicate that the station has successfully received all traffic sent on the multicast circuit.
5 ("UNL_NAK")	Sent by a participant in a multicast circuit link in response to a TM_TERM PDU with Reason = UNLINK, to indicate that the station is still present in the multicast circuit, but has not received all traffic sent on the multicast circuit successfully
6 ("SUSPEND")	Suspends the current multicast circuit link while the link initiator repeats the robust LSU multicast call in order to retrieve as many as possible of the multicast group members that were found to be absent in the most recent roll call. Stations receiving the TM_TERM PDU with Reason = SUSPEND are expected to remain on the traffic channel for a time period sufficient to allow the link initiator to complete the robust LSU multicast call, return to the traffic channel, and send a TM_REQ PDU to start another roll call.
7–63	Reserved.

forward direction, the first xDL_DATA PDU is sent at a fixed time interval (known a priori to both stations) following the transmission of the TM_REQUEST PDU. Likewise, in the reverse direction, the first xDL_ACK PDU is sent at a fixed time interval following the transmission of the TM_CONFIRM PDU.

5.5 Data Transfer

The role of the 3G data link protocols is to convey a data payload, or datagram, from a source to a destination over an HF link as efficiently as possible. The 3G data link protocols are focused on error-free data delivery because the serial tone waveforms of MIL-STD-188-110 and the 2G unacknowledged data link protocols are available for use where delivery with errors can be tolerated. The 3G protocols all use automatic repeat request (ARQ) techniques to deliver the payload data without errors as far as possible. This requires two-way communications with the destination node reporting back to the origination node that an error-free message has been received. The general approach utilized is to divide the data message into a number of frames or packets, protecting each frame with a strong CRC code such that, upon reception, the entire frame can be checked for errors. If errors are detected, a request for retransmission is sent from the destination to the origination station to repeat the last transmission. If the frame is received error-free, a request to continue to the next frame or packet is sent from the destination to the origination station. A number of ARQ techniques exist, but they all practice variations on the above technique in order to convey data error free over a communications link.

5.5.1 3G Data Link Protocols: A New Approach

As discussed in Chapter 3, the HF channel and associated radio systems provide a number of challenges to the goal of error-free data transmission.

- HF propagation is a significant factor. The ALE system must identify a channel that provides adequate communications in both link directions.
- The variability of HF channel characteristics (including the received signal-to-noise ratio) makes the selection of on-air modem parameters difficult. If the selected modem waveform is insufficiently robust to tolerate the channel's received SNR, multipath spread, or Doppler spread, then the data will not be transferred error-free. If the modem mode selected is too conservative (relative to the HF channel conditions), the data will be transferred error-free, but at a lower transfer rate than could be sustained, reducing overall system performance.
- The link turnaround time of an HF system also plays a significant role in the performance of a data link protocol, as was discussed in Chapter 3. HF data modems use coding and interleaving to address the short-term fading and interference of the HF channel. This results in a processing delay between the reception of the waveform and the availability of the data bits that need to be checked by the CRC code. Additionally, there is a fixed time required to transition the HF radio equipment from receive to transmit in order to send the ACK message back to the destination.
- Given the variability of the HF channel, the data link protocol must be able to withstand complete loss of transmissions, either the forward data transmission or the return ACK transmission. This situation is made worse by some data link protocols, notably the 2G protocols, which attempt to vary the on-air modem parameters of bit rate and interleaver depth to match the channel variation. As these parameters are changed, the on-air transmission durations can change. This can lead to protocol state confusion at either end of the link.

To address these issues, a new suite of ARQ data link protocols was developed as part of the 3G standards. These new protocols include a low-latency data link protocol LDL, a high-throughput data link HDL, and the still higher capacity protocol known as HDL+ (pronounced "HDL plus").

These new standards utilize the robust burst waveforms (described earlier in Section 5.2) to be used in the ACK signaling. These waveforms are specifically designed to be robust and short in duration, and they reduce the chances of missing an ACK signal.

Significantly, these new standards addressed the variability of the HF channel by applying the concepts of code combining or type II hybrid ARQ systems [10]. In these systems, a common on-air modulation is utilized; LDL uses a variation of the Walsh orthogonal modulation, providing a top uncoded bit rate of 600 bps, while HDL uses a variation of the equalized 4800 bps higher rate HF data modem. HDL+ (a candidate for NATO standardization) is a variation of these approaches that uses the STANAG 4539 waveforms (see Chapter 3, Section 3.2.3), which provide data rates of up to 12,800 bps (uncoded).

In these type II hybrid ARQ systems, each forward transmission carries a different phase or output of the convolutional forward error correction code. With each reception, soft decision metrics computed by the modem equalizer are combined with those from previous receptions and decoding is attempted. In a sense, it is not the on-air bit rate that is being adapted to try and match what the current channel will support, but the effective forward error correction code rate. For example, if the initial reception is decoded error-free, the effective code rate would be 1. If two receptions are required, the effective code rate would be 1/2. If three receptions are required, the effective code rate would be 1/3, and so on. In this approach, all received information is used to help decode the packet.

In the more traditional approaches, if the packet was received with errors, all the information would be dropped and another attempt at reception would be tried, making no use of the previously received data. For example, a packet of *X* bytes may have been received with only one or two errors. Why disregard all of this information?

There is another benefit to this approach: because the modulation is not adapted for each forward transmission, all transmission times are the same, and the two stations know when to expect the beginning of a reception. This greatly reduces the protocol confusion that can arise in 2G data link protocols (Chapter 3) when portions of a transmission are lost to fading.

The combination of robust burst signaling, the performance improvements of type II ARQ methods, and the corresponding fixed transmission times have resulted in a set of 3G data link protocols that provide significant performance gains over the performance achievable with the 2G adaptive data rate protocols.

There is an additional performance gain associated with the type II ARQ approach. The method of effectively adapting the forward error correction code rate results in a system that can much more easily track the variations of the HF channel, especially when compared with the 2G data link protocols (see Section 5.5.5.1).

5.5.2 LDL: Low-Latency Data Link Protocol

The low-latency data link protocol (LDL) is a stop-and-wait ARQ protocol (see Chapter 3, Section 3.3.1) that provides reliable acknowledged point-to-point delivery of datagrams from a transmitting station to a receiving station across an already-established HF link. The datagram passed to the LDL protocol entity for delivery is an ordered sequence of up to 16,384,000 8-bit data bytes (octets). The

LDL protocol provides better performance than does the high-throughput data link protocol for all datagram lengths under fair to very poor HF channel conditions, and under all channel conditions for short datagrams.

Data transfer by LDL begins after traffic management has configured the data link connection, including the fact that LDL will be used (as opposed to HDL or another data link protocol), and the precise time synchronization of data link transmissions.

In an LDL data transfer, the sending station and the receiving station alternate transmissions in the manner depicted in Figure 5.37, the sending station transmitting LDL_DATA PDUs containing payload data packets, and the receiving station transmitting LDL_ACK PDUs each containing an acknowledgment of whether or not the data packet in the preceding LDL_DATA PDU was received without error. If either station fails to receive a PDU at the expected time, it sends its own next outgoing PDU at the same time, as if the incoming PDU had been received successfully. The times at which the burst waveforms conveying LDL_DATA, LDL_ACK, and LDL_EOM PDUs may be transmitted are determined precisely by the initial data link timing established during the traffic setup phase. These key features significantly add to the robustness of an LDL datagram transfer under challenging HF channel conditions.

The end of a data transfer is reached when the sending station has transmitted LDL_DATA PDUs containing all of the payload data in the delivered datagram, and the receiving station has received these data without errors and has acknowledged their successful delivery. When the sending station receives an LDL_ACK PDU indicating that the entire contents of the datagram have been delivered successfully, it sends an LDL_EOM PDU repeated as many times as possible within the duration of an LDL_DATA PDU, starting at the time at which it would have otherwise transmitted the next LDL_DATA PDU; this indicates to the receiving station that the data transfer will be terminated. This link termination scenario is depicted in Figure 5.38.

5.5.2.1 LDL PDUs

Figure 5.39 depicts the format and contents of the LDL PDUs. Each LDL_DATA PDU carries, a single data packet composed of payload data (N × 32 *bytes* (octets) of user data, where *N* ranges from 1 to 16) followed by a 17-bit sequence number and an 8-bit control field (presently unused) added by the LDL protocol. During the traffic setup phase, the user process determines the number of data bytes per LDL_DATA PDU so as to deliver the user data efficiently, shortening the LDL_DATA PDU whenever the entire datagram is short enough to fit within the shortened

Figure 5.37 LDL data transfer overview.

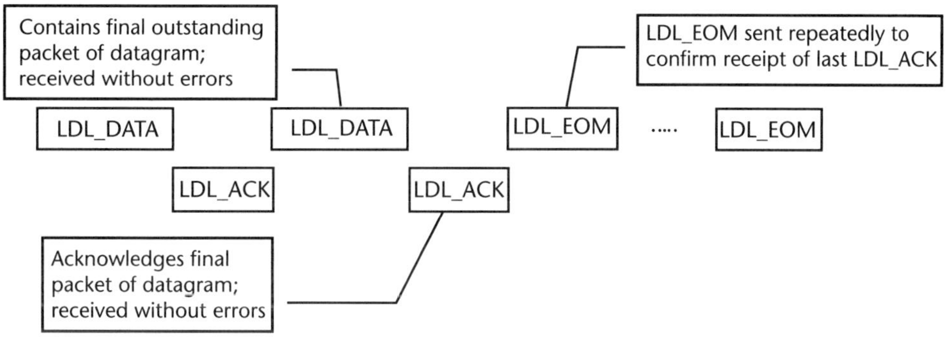

Figure 5.38 LDL link termination scenario overview.

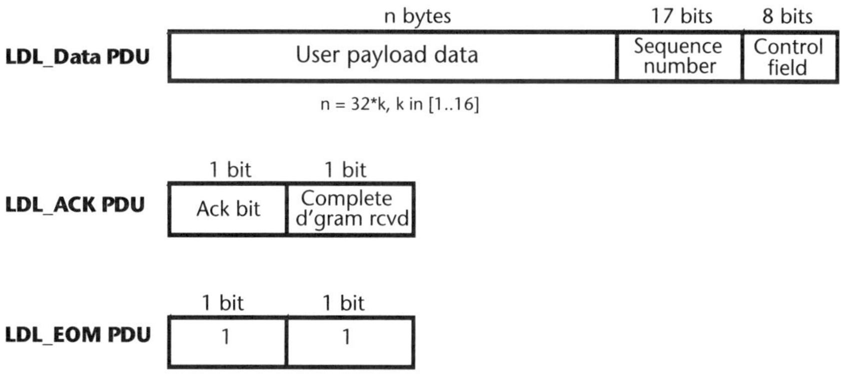

Figure 5.39 LDL PDUs.

PDU. Once it is determined, the number of data bytes per LDL_DATA PDU for the current datagram delivery is communicated to the receiving station in the traffic setup sequence. Thereafter, every transmitted LDL_DATA PDU contains the same number of data bytes until the entire datagram has been delivered.

The LDL_ACK PDU is used to convey data acknowledgments from the receiving station to the sending station. Each LDL_ACK PDU contains acknowledgment for the immediately preceding LDL_DATA PDU sent in the opposite direction; the single bit in the Ack bit field acknowledges the single data packet in the LDL_DATA PDU. The "Complete datagram rcvd" bit is set when the receiving station determines that it has received all of the contents of the datagram without errors, so that the data link transfer can be ended. The LDL_ACK PDU is transmitted using the very robust BW4 waveform. Due to the robustness of the waveform, no CRC is included in the PDU.

The LDL_EOM PDU is transmitted in the forward direction, in place of an LDL_DATA PDU, when the sending station receives an error-free LDL_ACK PDU indicating that the entire user datagram has been delivered to the receiving station without errors. This PDU is also transmitted using the BW4 waveform. LDL_EOM PDUs are distinguished from LDL_ACK PDUs by context: any BW4 transmission in the *forward* direction of an LDL transfer is an LDL_EOM PDU.

5.5.2.2 LDL ARQ Processing

Figure 5.40 depicts the manner in which an LDL_DATA PDU is incorporated into a BW3 burst transmission. The LDL_DATA PDU is extended by appending to it a 32-bit CRC, followed by an encoder flush sequence consisting of seven zero bits. The resulting data sequence is FEC-encoded using a ½-rate convolutional encoder. The encoder produces two output bits, $Bitout_0$ and $Bitout_1$, for each input bit. As each packet is encoded, the bits from each encoder output are accumulated into a block, resulting in two blocks of coded bits, $EBlk_0$ and $EBlk_1$. Each time a data packet is transmitted in an LDL_DATA PDU, only one of the two blocks of coded bits from the packet is transmitted, starting with $EBlk_0$ the first time. (The original contents of the packet can be recovered from any single block of coded bits that is received without errors.) Each time the data packet cannot be decoded without errors and must be retransmitted, a different block of coded bits is transmitted; blocks are transmitted in the order $EBlk_0$, $EBlk_1$, $EBlk_0$, $EBlk_1$, ... The transmission of different blocks of coded bits for each packet, in successive transmissions of the packet, provides additional information that can be used in decoding the packet.

The sequence of coded bits is interleaved using a convolutional block interleaver similar to that of MIL-STD-188-110C. The interleaved bit sequence is then modulated using a modulation process similar to that of the MIL-STD-188-110C serial tone waveform at 75 bits per second (described in Chapter 3, Section 3.2.2.2). This results in a sequence of 16-ary orthogonal Walsh frames, each composed of 16 PSK symbols, with each 16-ary Walsh frame representing the values of four coded bits fetched from the interleaver.

An acquisition preamble of 640 8-ary PSK symbols is prepended to the beginning of the Walsh frame sequence. This preamble is used for initial channel estimation, frequency offset estimation, and fine time alignment synchronization in order to make the most use of any present multipath. The preamble need not be used for coarse synchronization, since the time of arrival of each BW3 transmission is known once the traffic setup handshake is used to establish data link timing.

5.5.3 High-Throughput Data Link Protocol

The high-throughput data link protocol (HDL) is a selective repeat ARQ protocol, used to provide acknowledged point-to-point delivery of datagrams from a

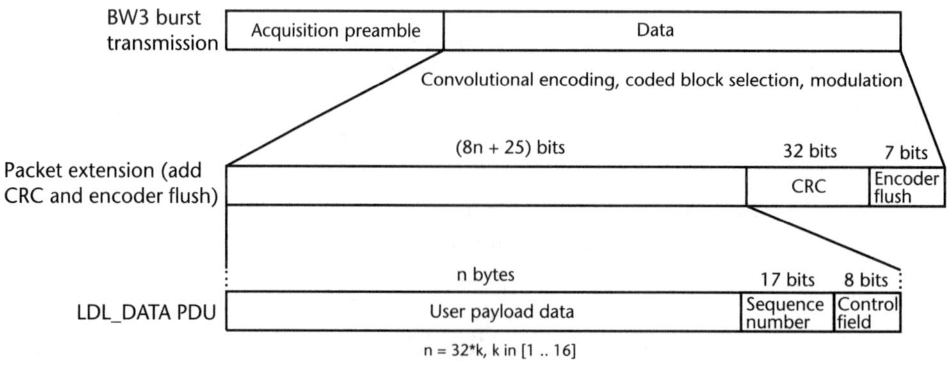

Figure 5.40 BW3 encoding and modulation of LDL_DATA PDU.

transmitting station to a receiving station across an already-established HF link. The datagram passed to HDL for delivery is an ordered sequence of up to 7,634,944 8-bit data bytes (octets). The HDL protocol is best-suited to delivering relatively large datagrams under good-to-fair HF channel conditions. By contrast, the low-latency data link protocol described above provides better performance for all datagram lengths under fair-to-very poor HF channel conditions, and under all channel conditions for short datagrams.

Data transfer by HDL begins after the stations have already established the data link connection in the traffic setup phase. This determines that HDL will be used (as opposed to LDL or some other mechanism), the number of data packets to be sent in each HDL_DATA PDU, and the precise time synchronization of data link transmissions.

In an HDL data transfer, the sending station and the receiving station alternate transmissions in the manner depicted in Figure 5.41; the sending station transmitting HDL_DATA PDUs containing payload data packets, and the receiving station transmitting HDL_ACK PDUs containing acknowledgments of the data packets received without errors in the preceding HDL_DATA PDU. If either station fails to receive a PDU at the expected time, it sends its own next outgoing PDU at the same time as if the incoming PDU had been received successfully. The times at which the burst waveforms conveying HDL_DATA, HDL_ACK and HDL_EOM PDUs may be transmitted are determined precisely by the initial data link timing established during the link setup (in FLSU) or traffic setup (in RLSU) phase.

The end of a data transfer is reached when the sending station has transmitted HDL_DATA PDUs containing all of the payload data in the delivered datagram, and the receiving station has received these data without errors and has acknowledged their successful delivery. When the sending station receives an HDL_ACK PDU indicating that the entire contents of the datagram have been delivered successfully, it sends an HDL_EOM PDU repeated as many times as possible within the duration of an HDL_DATA PDU, starting at the time at which it would have otherwise transmitted the next HDL_DATA PDU, to indicate to the receiving station that the data transfer will be terminated. This link termination scenario is depicted in Figure 5.42.

5.5.3.1 *HDL PDUs*

Figure 5.43 depicts the format and contents of the HDL PDUs. Each HDL_DATA PDU is a sequence of 24, 12, 6, or 3 data packets, in which each packet is composed of 1881 bits of payload data (1864 bits of user data, plus a 17-bit sequence number added by the protocol). During the link/traffic setup phase, the session manager process determines the number of data packets per HDL_DATA PDU so

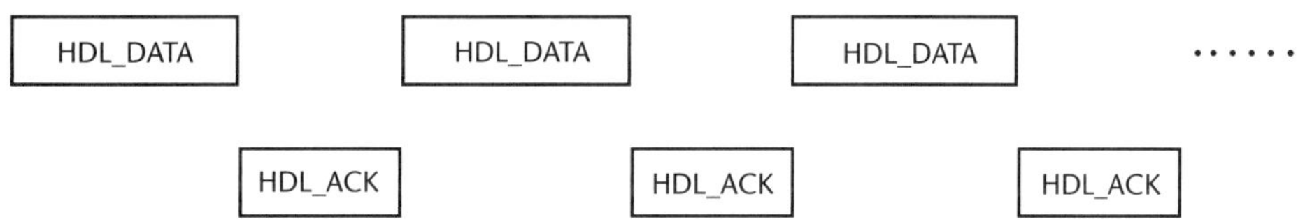

Figure 5.41 HDL data transfer overview.

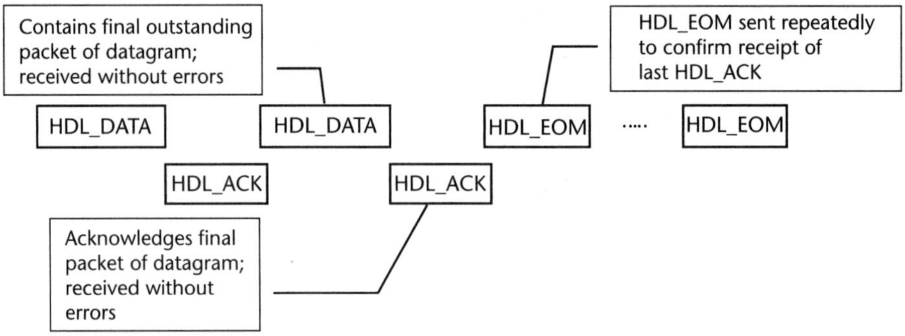

Figure 5.42 HDL link termination scenario overview.

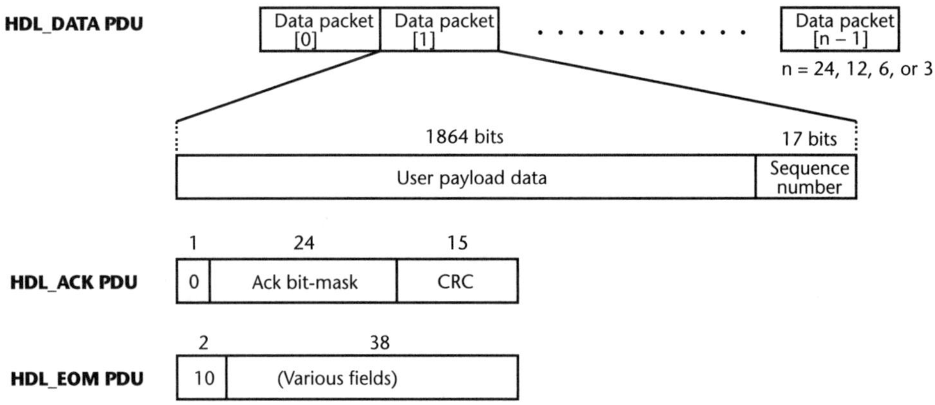

Figure 5.43 HDL PDUs.

as to deliver the user data efficiently, shortening the HDL_DATA PDU whenever the entire datagram is short enough to fit within the shortened PDU. Once it is determined, the number of data packets per HDL_DATA PDU for the current datagram delivery is communicated to the receiving station in the link/traffic setup sequence. Thereafter, every HDL_DATA PDU contains the same number of data packets until the entire datagram has been delivered. The BW2 waveform is used to transmit each HDL_DATA PDU; further description of BW2 processing is provided in Section 5.2.4.

The HDL_ACK PDU is used to convey data acknowledgments from the receiving station to the sending station. Each HDL_ACK PDU contains acknowledgments for the immediately preceding HDL_DATA PDU sent in the opposite direction; each bit in the *Ack bit-mask field* acknowledges a single corresponding data packet from the HDL_DATA PDU. The HDL_ACK PDU contents are protected by a 16-bit CRC.

The HDL_EOM PDU is transmitted in the forward direction, in place of an HDL_DATA PDU, when the sending station receives an error-free HDL_ACK PDU indicating that the entire user datagram has been delivered to the receiving station without errors.

BW1 is used to transmit both the HDL_ACK and HDL_EOM PDUs. The marker bits at the beginning of each PDU are used to distinguish the two kinds of PDUs.

5.5.3.2 HDL ARQ Processing

Figure 5.44 depicts the manner in which an HDL_DATA PDU is incorporated into a BW2 burst transmission. Each data packet in the HDL_DATA PDU is extended by appending to it a 32-bit CRC, followed by an encoder flush sequence consisting of seven zero bits. The resulting sequence of extended packets is FEC-encoded and uses a ¼-rate convolutional encoder. The encoder produces four output bits, $Bitout_0$ .. $Bitout_3$, for each input bit. As each packet is encoded, the bits from each encoder output are accumulated into a block, resulting in four blocks of coded bits, $EBlk_0$ .. $EBlk_3$. Each time a data packet is transmitted in an HDL_DATA PDU, only one of the four blocks of coded bits from the packet is transmitted, starting with $EBlk_0$ the first time. (The original contents of the packet can be recovered from any single block of coded bits that is received without errors.) Each time the data packet cannot be decoded without errors and must be retransmitted, a different block of coded bits is transmitted; blocks are transmitted in the order $EBlk_0$, $EBlk_1$, $EBlk_2$, $EBlk_3$, $EBlk_0$, ... The transmission of different blocks of coded bits for each packet, in successive transmissions of the packet, provides additional information that can be used in decoding the packet.

The sequence of coded bits is modulated using a process similar to that of the 110A serial tone waveform at 4800 bits per second. This results in a sequence of unknown/known symbol frames, each consisting of 32 unknown symbols (8-ary PSK symbols carrying three bits each, Gray-coded), followed by 16 known symbols. A TLC/AGC guard sequence and a sequence of 64 known symbols used for initial equalizer training are prepended to the beginning of the frame sequence.

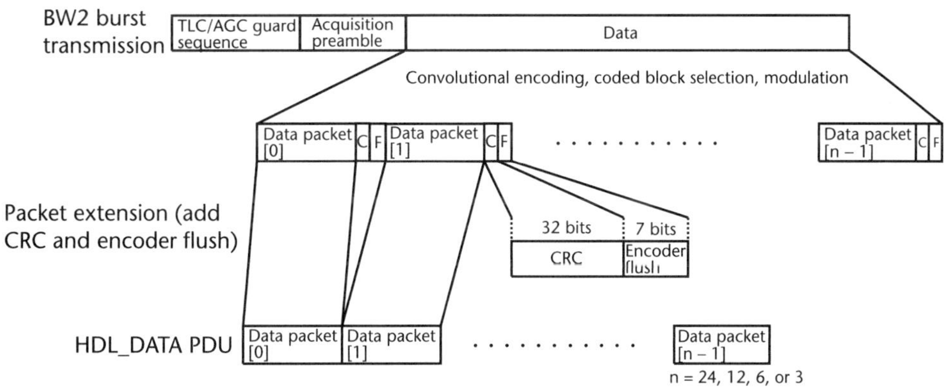

Figure 5.44 BW2 encoding and modulation of HDL_DATA PDU.

Note that no acquisition preamble is required in the BW2 waveform, since the precise time of arrival of each BW2 transmission is known once the link/traffic setup handshake is used to establish data link timing.

5.5.4 HDL+ Data Link Protocol

The HDL+ data link protocol combines high data rate waveforms similar to those of STANAG 4539 or MIL-STD-188-110C Appendix C with incremental redundancy (Type II Hybrid-ARQ) techniques similar to those of the STANAG 4538 LDL and HDL protocols [11]. As a result, it achieves excellent data transfer throughput rates under a wide variety of conditions, and rates of up to 10,000 bits per second in a 3-kHz channel under high-SNR Gaussian noise conditions. HDL+ is designed to be incorporated into the STANAG 4538 protocol framework alongside the LDL and HDL protocols of the ratified STANAG 4538 Edition 1 [1]; the station initiating a data transfer announces in the initial link setup or traffic management handshake whether LDL, HDL, or HDL+ is to be used for the ensuing data transfer.

While it has been proposed for incorporation into STANAG 4538, HDL+ has not been standardized as of this writing, although it has been implemented and widely used in the radio equipment of one prominent manufacturer. For this reason, the discussion of HDL+ in this section will have less technical detail than is present in the sections for LDL and HDL.

5.5.4.1 Background and Motivation

With the advent of the higher data rate waveforms in MIL-STD-188-110B Appendix C and STANAG 4539, it became desirable to offer similar data rates and the resulting increased potential throughputs within the 3G protocol family as well. The 8-PSK signal constellation and 32/16 unknown/known frame format of the HDL forward transmission limits it to an effective data rate of 4800 bps even without coding, comparing unfavorably with the 9600 bps possible with the 64-QAM constellation of 110B. At the same time, the growing desire for TCP/IP delivery capabilities over HF (including TCP and UDP) focused attention on design features of the HDL protocol that led to inefficiencies in the delivery of IP traffic: the inflexible transmission format and the large amount of overhead in delivery of small payloads, which is especially burdensome for such items as TCP ACKs and the numerous exchanges of “chatty” application protocols such as SMTP. The design of HDL+ is specifically intended to remove these limitations while retaining the robustness and throughput advantages made possible by the 3G-style robust burst waveforms and type II hybrid-ARQ techniques used in LDL and HDL.

5.5.4.2 Design Overview

HDL+ is an adaptive protocol, using a variety of code rates and signal formats to try to achieve the best possible throughput under different channel conditions.

- In the highest-rate format, a 64-QAM signal constellation is combined with a half-rate $k = 7$ convolutional code; however, each transmission of a packet contains only one of the two encode phases of the code, making this an application of a code-combining incremental-redundancy technique [10].

- At lower rates, the same half-rate code is punctured to rate 3/4 and the coded symbols are modulated using signal constellations 64-QAM, 16-QAM, 8PSK, QPSK, or BPSK. Successive transmissions contain the same code symbols; the form of incremental redundancy employed is diversity combining rather than code combining [10].
- In the lowest-rate format, the code is used at rate 1/2 rather than being punctured to rate ¾, and the signal constellation used is BPSK.

The use of this assortment of signal constellations within HDL+ allows the protocol to be used beneficially over a wide range of channel conditions, SNR in particular. Each data link acknowledgment PDU contains the receiving node's estimates of the signal-to-noise ratio (SNR) and Doppler spread with which the preceding forward transmission was received, providing information the sender can use to adjust the signaling format of the next forward transmission accordingly.

In HDL, the use of a fixed forward transmission length through the entire duration of each data link transfer led to inefficiencies, particularly at the end of a transfer when a shorter forward transmission might have sufficed to convey the remainder of the payload datagram. HDL+ addresses this issue by making forward transmission lengths variable from 1 to 15 packets.

HDL+ uses two packet sizes: 280 and 568 bytes. The resulting forward transmission duration depends on the signaling format, and can be up to 64.8 seconds for the half-rate BPSK format with 568-byte packets. Typical 15-packet forward transmission durations range from 7 to 25 seconds.

The variable lengths and modulation formats of HDL+ forward transmissions make it necessary to include a header at the beginning of each HDL+ forward transmission, which was unnecessary in LDL and HDL. For this header, the HDL+ design includes the definition of a new robust burst waveform, BW6, having 51 bits of payload and an on-air duration of 386.67 ms. The BW6 header at the beginning of each forward transmission announces the number of packets and modulation format of the following payload section of the transmission. BW6 is also used for acknowledgments and other protocol signaling related to HDL+. Since the data rate of this burst is higher than that of the BW5 bursts used in the STANAG 4538 fast link setup (FLSU) protocol, the low-SNR usability of HDL+ is limited by the somewhat reduced robustness of this burst format. (For lower-SNR conditions, the LDL and HDL protocols remain available.) However, the minimal overhead of the HDL+ protocol design permits it to achieve a maximum throughput of over 10,000 bps under benign channel conditions, although the throughput achieved on-air is usually lower on typical skywave channels.

5.5.4.3 Status of HDL+

HDL+ was designed to be incorporated into Edition 2 of STANAG 4538, and was proposed for incorporation into that standard. However, a disagreement over intellectual property rights contained in the technology—a patent held by the Harris Corporation—has precluded standardization of this protocol. As a result, HDL+

has been available only in the Harris Corporation's HF radio products of the Falcon II® product family, where its use has been found to be quite beneficial. It is hoped that this impasse can be resolved to make the benefits of the HDL+ protocol more widely available.

5.5.5 3G Data Link Performance

Throughout their development, the performance of the 3G data link protocols has been extensively characterized under both simulated and real-world conditions. Because of the difficulty in obtaining repeatable results under real-world conditions, the majority of this testing has been done under simulated channel conditions—specifically those of the Watterson model [14]. The most common simulated channel conditions have been the Gaussian and Mid-latitude disturbed channel conditions defined by ITU-R Rec. F.1487 [5]. In modem design, these two channel conditions are often considered adequate to validate the design concept or an implementation of a new waveform; so it is natural to want to use them in the same way in validating data link protocol design concepts and implementation. However, this by itself isn't enough; it is necessary to supplement testing on simulated channels with testing on real-world channels at frequencies and link distances representative of those likely to be encountered in the operation of a delivered radio system, and preferably in a manner capable of yielding informative side-by-side performance comparisons between different protocols or implementations. This is done so that (for instance) the performance of the 3G data link protocols can be compared with that of 2G data link protocols, such as FED-STD-1052 or STANAG 5066. In the performance testing reported below, this has often been achieved by testing one protocol, then another in an alternating fashion while holding frequency, antenna characteristics, power level, and so on constant.

In keeping with the widespread and growing use of HF for data message delivery such as HF E-mail, many of the performance testing scenarios have focused on delivery of messages: finite byte sequences that are typically 500, 5000, or 50k bytes in length. The performance measure used in reporting the results of these tests is typically throughput: delivered message size in bits divided by the delivery time in seconds. In doing this calculation, it is important to define precisely what one means by delivery time: whether it includes the initial link set-up or traffic set-up, or the one or more termination PDUs with which the data link transfer is concluded.

More recently, delivery of IP packet traffic has been increasing in importance, necessitating the development of a new suite of tests aimed at determining the 3G protocols' performance as bearers for IP traffic, and the impact of these performance levels on the functioning of application-layer protocols and the applications themselves.

5.5.5.1 LDL and HDL Performance

Data link protocol performance is typically defined and measured in terms of the average throughput in bits per second. The throughput achieved is dependent on many factors, including HF channel conditions, both short-term and long-term, as well as datagram size. Protocol parameters (which may be selected by the user or automatically adapted) can also play a large role in determining throughput. These parameters may include modem transmission rates, frame or packet size, link turnaround times, and so on.

Data link performance for the HDL and LDL protocols are presented in this section. Also presented in this section, for comparison, is the measured performance of a second generation adaptive data rate data link protocol, the United States Federal Standard 1052 data link protocol (1052) [15]. Federal Standard 1052 uses the U.S. MIL-STD-188-110A serial tone modem waveform. The autobaud capability of thc modem is used extensively as the protocol adapts the data rate and interleaver settings to the HF channel conditions.

The throughput rates presented in the following figures account for the entire time spent on the traffic frequency after the completion of a successful link setup (RLSU), including the time for the TM handshake and the time for the transfer of the message. For HDL and LDL, the traffic setup times are based on the BW1 handshake. In the case of the 1052, the data link's own handshake timing is included instead of the BW1 handshake timing.

Figures 5.45 through 5.50 present measured throughput performances under simulated channel conditions for HDL and LDL, and compare their performance to measured 1052 performance for 50-byte, 500-byte, and 50-Kbyte files. Each data link protocol's performance is given for the additive Gaussian noise (AWGN) and mid-latitude disturbed (MLD) HF channel conditions [5]. AWGN refers to a single nonfading path with additive white Gaussian noise. Mid-latitude disturbed refers to dual fading paths, separated by 2 ms, each path with a two-sigma fading bandwidth of 1 Hz.

In comparing the HDL and LDL throughput performance curves against 1052, one must be careful to understand some of the basic differences between the two data link protocol techniques. The throughput curves for 1052 do not contain a two-way handshake between the two stations in transfers of smaller files (i.e., 50 and 500 bytes); instead, 1052 sends a single one-way herald at the beginning of the data transfer. The handshake in the front of the data transfer can dominate throughputs for small files.

For larger files, 1052 uses a two-way call-response handshake and link termination as part of the file transfer, and therefore allows for a better comparison

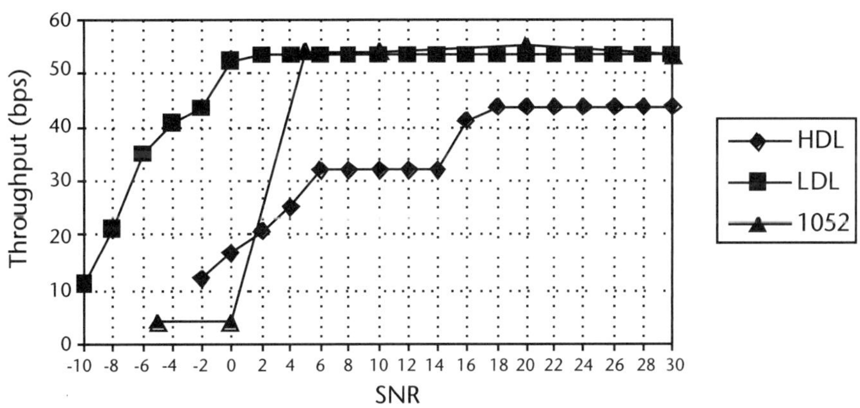

Figure 5.45 AWGN channel, 50-byte message. (Source: [8].)

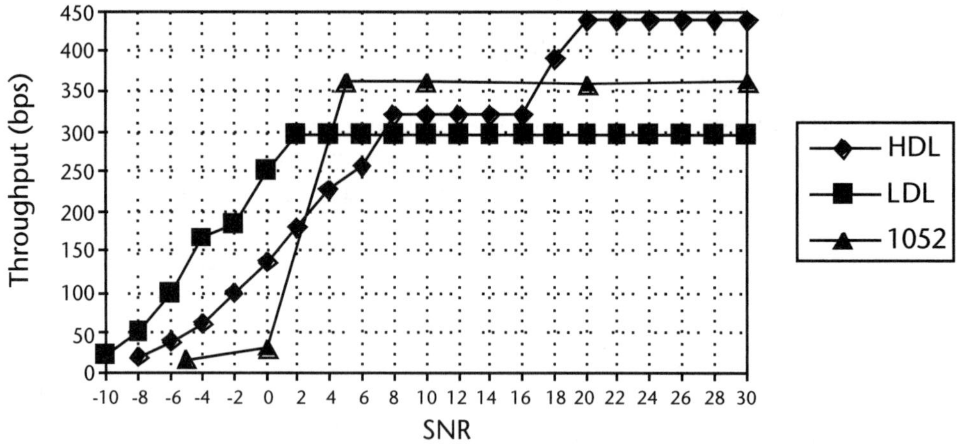

Figure 5.46 AWGN channel, 500-byte message. (Source: [8].)

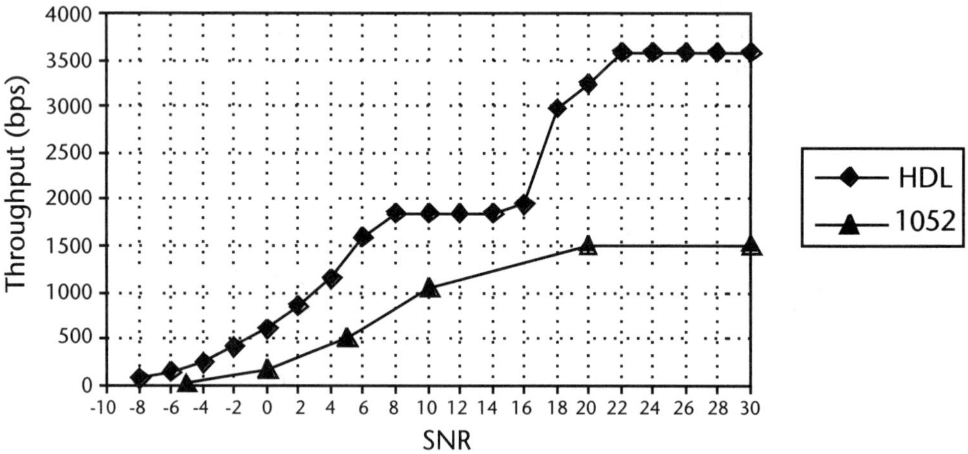

Figure 5.47 AWGN channel, 50-Kbyte message. (Source: [8].)

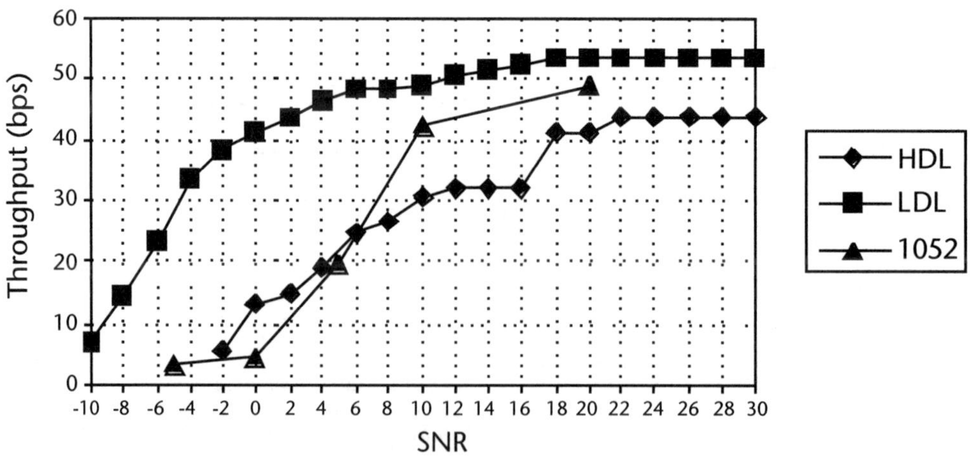

Figure 5.48 MLD channel, 50-byte message. (Source: [8].)

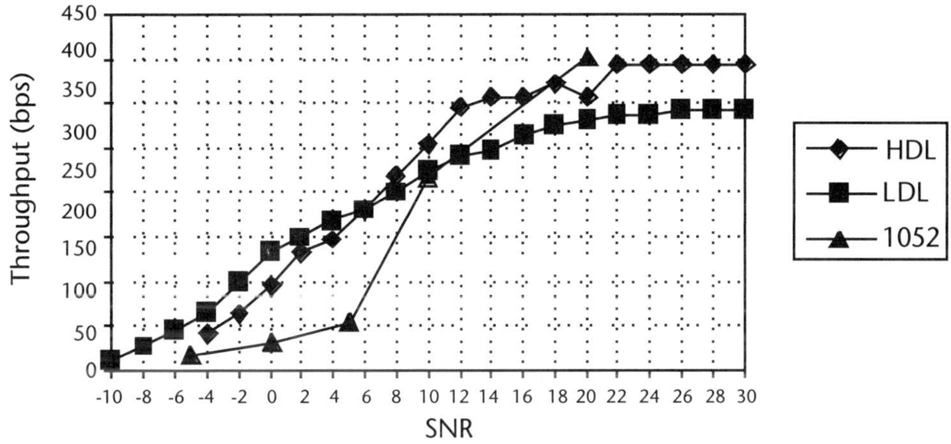

Figure 5.49 MLD channel, 500-byte message. (Source: [8].)

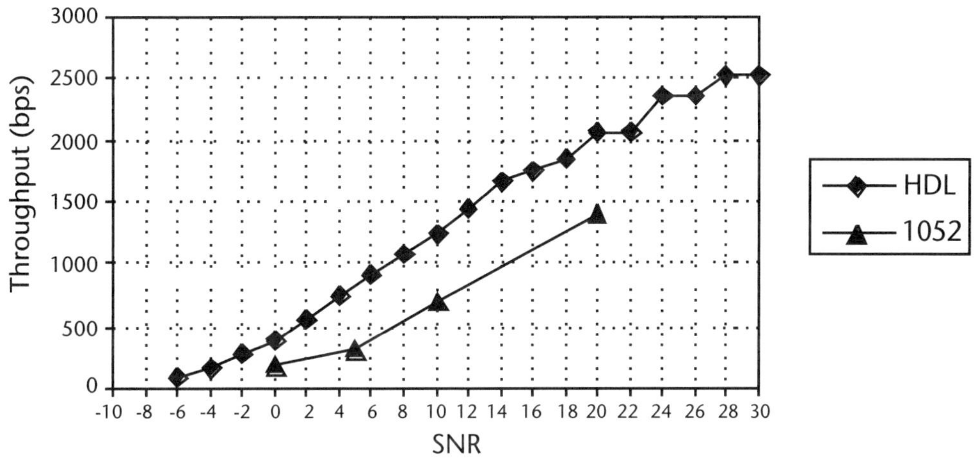

Figure 5.50 MLD channel, 50-Kbyte message. (Source: [8].)

between techniques. Also note that 1052's user-selected forward bit-rate setting can bias its indicated throughput for smaller files. A high initial bit-rate setting helps for high SNR at the cost of reduced throughput for low SNR conditions. The 1052 data presented here uses 1200 bps as the initial forward bit-rate setting. Finally, overall performance of either system is influenced by the call setup and traffic setup mechanisms, as well as by the data link protocol. It is hard to compare the performance of the data link protocols in isolation because the data link protocol is not always the limiting factor in the performance of the system as a whole.

In comparing HDL to LDL, some interesting observations can be made. HDL has been optimized for higher throughputs for fair-to-good channel conditions. LDL has been optimized for better operation under severe to fair channel conditions through the choice of its underlying waveform. LDL's orthogonal Walsh signaling allows for better throughput performance at lower signal-to-noise ratios than does HDL's 8-ary PSK signaling. Also, LDL performs better for small message sizes because it incurs less overhead than HDL at these sizes. As a selective repeat ARQ protocol, HDL incurs less protocol overhead for larger files because of its better ratio of forward- to back-channel transmission times. Therefore, HDL is more efficient at the high end of the curves for larger file sizes.

It is important to see that in all of the conditions presented, either HDL or LDL provides throughput performance at least roughly equal to that of 1052; in many conditions, the performance of HDL or LDL is dramatically superior. In many conditions, HDL or LDL achieves throughput performance equal to that of FS-1052 at much lower SNR. This fact allows the delivery of equivalent 1052 throughput performance at a substantial reduction in radio transmit power. Additionally, for good SNR conditions, HDL can deliver substantially higher throughputs than can 1052.

5.5.5.2 HDL+ Performance

As the name suggests, HDL+ was originally conceived in part as an improvement over HDL, using new signaling formats inspired by those of the MIL-STD-188-110B Appendix C waveforms to provide much higher delivery throughput. For this reason, many of the throughput performance data for HDL+ have been presented in terms of comparisons with HDL.

This section describes the simulation results for HDL and HDL+ under various test conditions originally presented by Chamberlain [11]. Performance results are shown for a software simulation model of the proposed data link protocol enhancement for STANAG 4538. All STANAG 4538 results measure the protocol's throughput, including a two-way channel setup and link termination. The HDL+ throughput measurements do not include the time devoted to link setup, as it is not required for every exchange.

The simulations use a software model of an HF channel simulator, which had been originally validated for use by the NATO technical working group on robust HF waveforms. This simulator is implemented per the recommendations outlined in Furman and Nieto [12].

Results are presented for two different HF channel conditions. AWGN refers to a single non-fading path with additive white Gaussian noise. ITU-MLD refers to a mid-latitude disturbed channel condition as defined by ITU-R Recommendation F.1487, with dual fading paths, separated by 2 ms, each with a fading bandwidth of 1 Hz. All simulations include the effects of 3 kHz transmit and receive radio filters.

These simulations utilize a maximum forward transmission of 128 packets, as opposed to the 15 packets mentioned previously. This represents an upper bound on the performance obtained by the HDL+ protocol. Limiting the forward transmission to 15 packets has little to no detrimental effect on the 5000-byte message payload performance and only a minor impact on the 50,000-byte performance. The benefit of this reduction is a significant decrease in required memory and associated power savings for tactical implementations. Additionally, the protocol is more responsive to external factors, such as interruptions.

Figure 5.51 shows the results of the simulated performance of HDL and HDL+ for the transfer of a 5000-byte message payload under AWGN channel conditions. Here, we can see that the HDL+ transfer has superior data throughput throughout the SNR range of 10 dB to 30 dB with HDL+ achieving significant gains past the 18 dB point where HDL performance reaches its maximum.

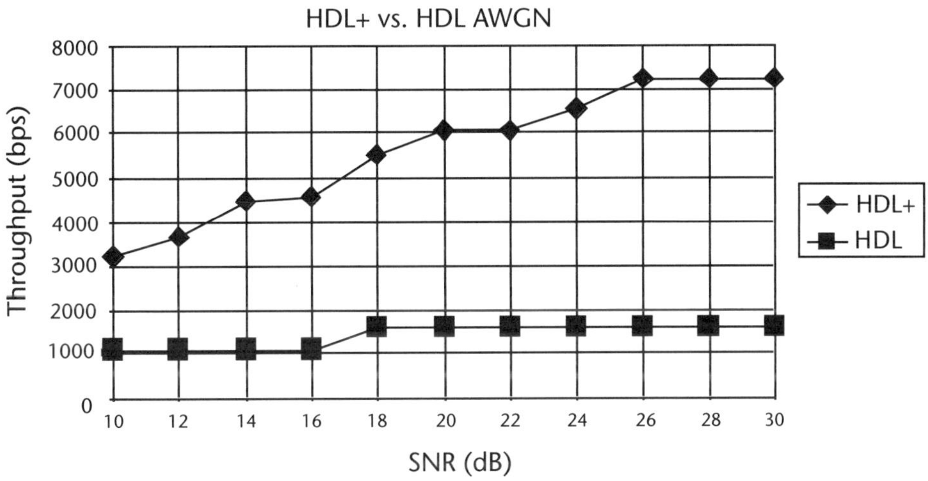

Figure 5.51 HDL / HDL+ Comparison 5000 byte AWGN. (© 2003 IET. Reprinted with permission from [11].)

Figure 5.52 displays a similar comparison for the case of a 5000-byte message transfer for the ITU-MLD channel. Once again we see a significant gain in throughput offered by HDL+.

Figures 5.53 and 5.54 display the throughput comparison for HDL and HDL+ for the channel conditions of AWGN and ITU-MLD for a 50,000-byte message payload transfer.

In examining the 50,000-byte message payload results, we see that, under AWGN channel conditions, HDL+ throughput approaches 12,000 bits per second (bps), and under ITU-MLD channel conditions throughputs approach 9,000 bps.

These results suffice to show the considerable increase in performance made possible by the design features of HDL+ that differentiate it from HDL. However, the data presented here actually understate the performance advantage HDL+ is

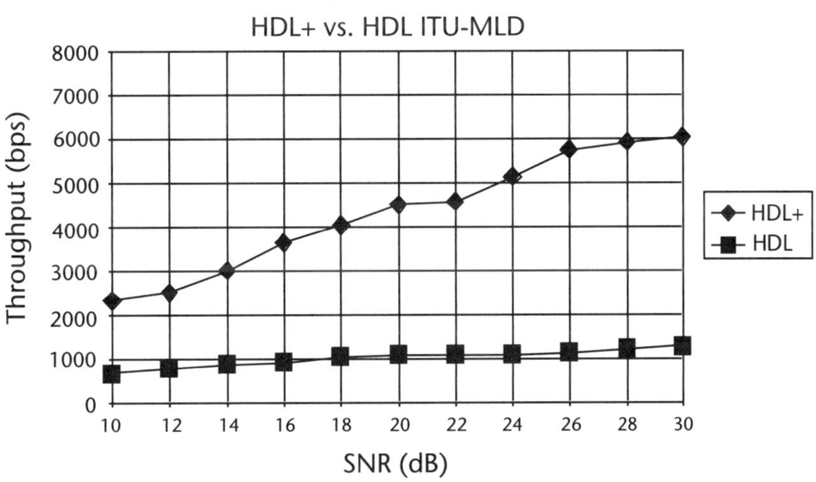

Figure 5.52 HDL / HDL+ comparison 5000-byte ITU-MLD. (© 2003 IET. Reprinted with permission from [11].)

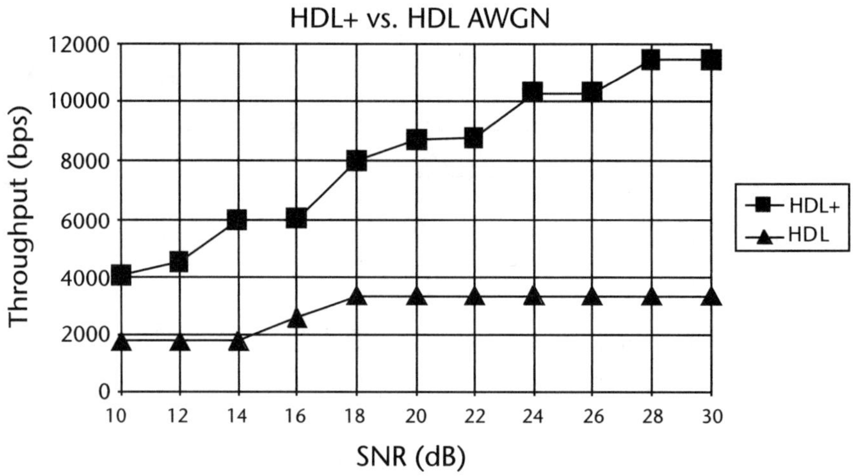

Figure 5.53 HDL / HDL+ comparison 50,000 byte AWGN. (© 2003 IET. Reprinted with permission from [11].)

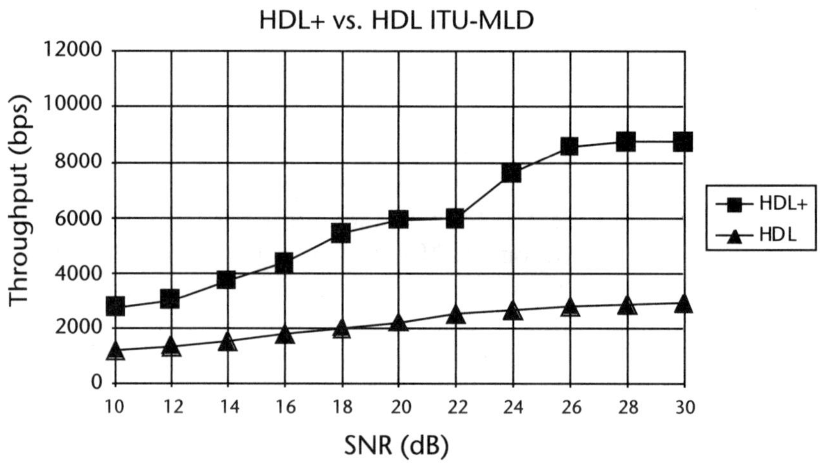

Figure 5.54 HDL / HDL+ comparison 50,000 byte ITU-MLD. (© 2003 IET. Reprinted with permission from [11].)

likely to exhibit on real-world HF links, especially by comparison with a 2G data link protocol such as STANAG 5066. Batts et al. [13] have shown that sky wave ionospheric paths in the HF ranges exhibit variations in SNR over medium- and long-term periods of a few seconds to a couple of minutes, which are not reflected in the Watterson channel model (at least, not as commonly used), but do significantly impact data link protocol performance; their measurements of this SNR variation and methods for modeling it are described in Chapter 2. Here, though, we can see how these intermediate and long-term variation phenomena can affect the performance comparisons between two data link protocols, a 2G protocol (STANAG 5066) and a 3G protocol (HDL+).

Figure 5.55 provides a throughput comparison between HDL+ and STANAG 5066 under Gaussian noise channel conditions. It can be seen that the throughput

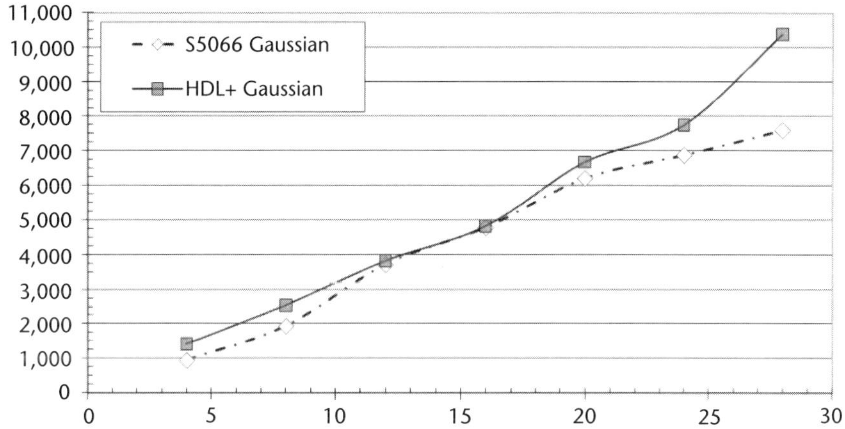

Figure 5.55 HDL+ versus S5066 throughput, Gaussian channel condition. (© 2007 IEEE. Reprinted with permission from [13].)

advantage of HDL+ is quite modest in this case, seeming to vanish entirely at an SNR of around 16 dB. On a mid-latitude disturbed channel with fading and multipath as in Figure 5.56, the performance advantage of HDL+ is more evident: a fairly consistent 2 to 3 dB or more. Figure 5.57 adds the intermediate- and long-term SNR variation characteristics to the simulated channel behavior, based on measured characteristics of the skywave path from Rochester, New York to Melbourne, Florida; here, the performance advantage of HDL+ is quite pronounced.

To gain a better understanding of the individual effects of the intermediate-term variation (ITV) and long-term variation (LTV) channel variation processes, additional testing was performed at an average signal to noise ratio of +20dB. Here, the SNR standard deviation in dB was adjusted individually for both ITV and LTV, each covering the range from 0 dB to 6 dB in steps of 2 dB. These 16 average throughput values were then plotted in the three-dimensional bar charts shown in Figures 5.58 and 5.59. The throughput data in both of these plots have

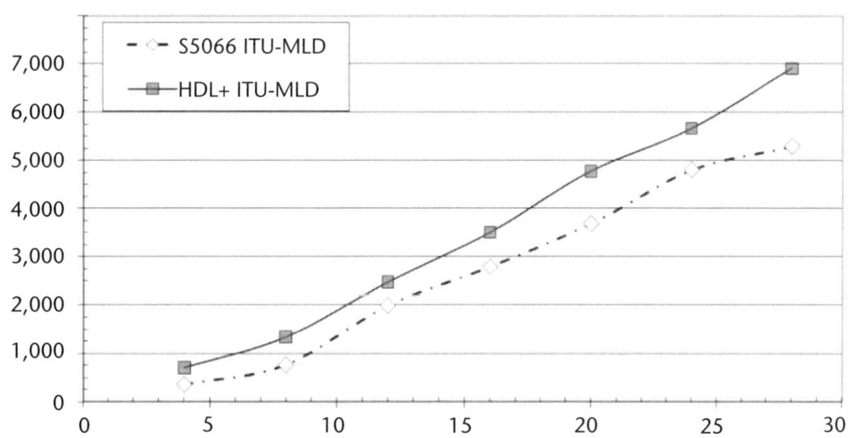

Figure 5.56 HDL+ versus S5066 throughput, ITU-MLD channel condition. (© 2007 IEEE. Reprinted with permission from [13].)

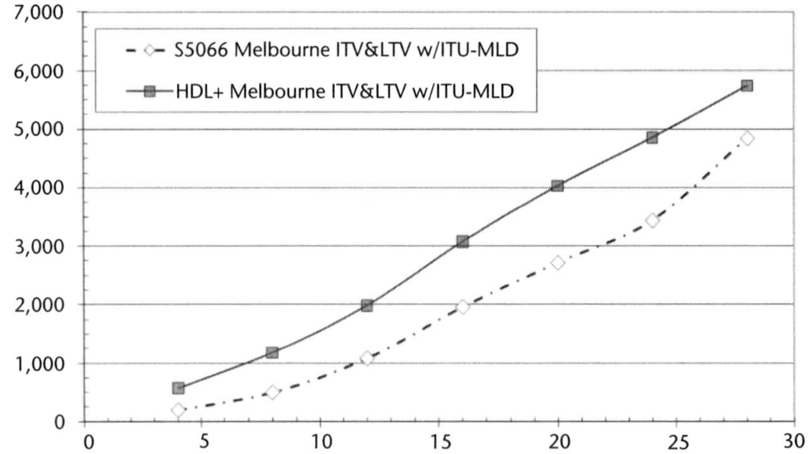

Figure 5.57 HDL+ versus S5066 throughput, ITU-MLD channel condition with ITV and LTV calibrated to Melbourne 070223 data set. (© 2007 IEEE. Reprinted with permission from [13].)

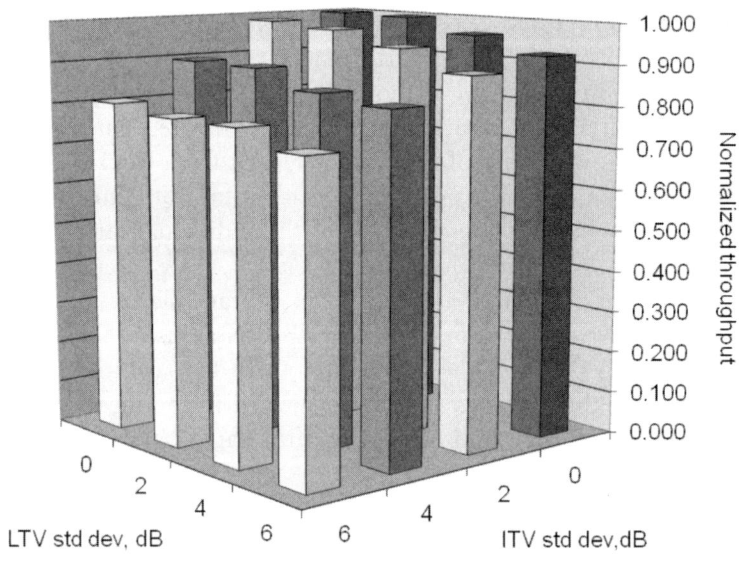

Figure 5.58 HDL+ normalized throughput, ITU-R MLD, +20 dB mean SNR. (© 2007 IEEE. Reprinted with permission from [13].)

been normalized to the highest achieved by HDL+ for the case of just the ITU-MLD channel, with no ITV or LTV.

Comparing Figure 5.58 to Figure 5.59 illustrates the relative insensitivity of HDL+ throughput performance to either ITV or LTV. This highlights the ability of the type II hybrid-ARQ protocol to effectively accommodate the changing channel conditions. The throughput does decrease as the ITV and LTV standard deviation values are increased, but even at the worst case of 6 dB ITV and 6 dB LTV, the normalized throughput is still above 70%. The data in Figure 5.56 indicate that S5066 achieves lower throughput than HDL+ for the case of no ITV and no LTV.

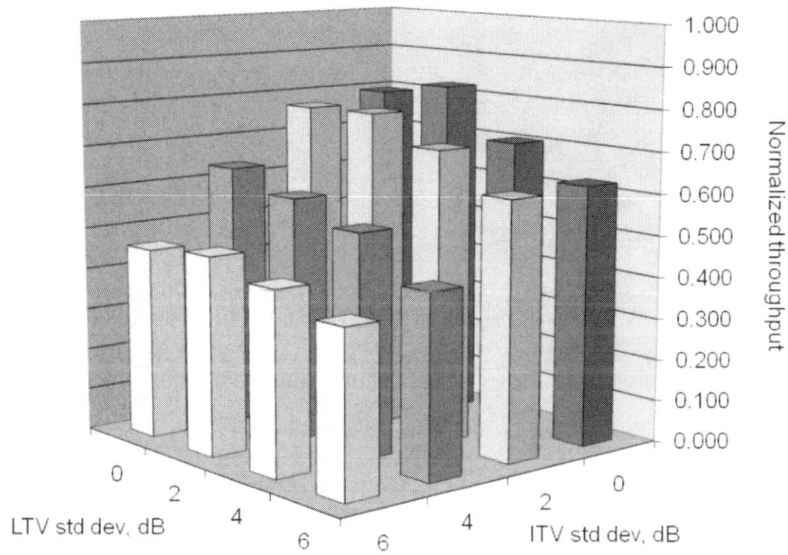

Figure 5.59 STANAG 5066 normalized throughput (vs HDL+), ITU-R MLD, +20 dB mean SNR. (© 2007 IEEE. Reprinted with permission from [13].)

Even from this lower starting point, S5066's throughput performance is further reduced by the addition of ITV and LTV, dropping to a worse case of about 39% and suffering a larger degradation than occurs for HDL+. Examining both Figure 5.58 and Figure 5.59 also demonstrates that both protocols are more susceptible to variations in ITV with its time constant of 5.2 seconds than they are to LTV with its time constant of 180 seconds. LTV, with its longer time constant, generates slowly changing channel variations that both protocols can accommodate reasonably well by adapting the data rate.

Figure 5.60 depicts the ratio of STANAG 5066 throughput to HDL+ throughput under differing amounts of channel quality variation. The much greater negative impact of channel quality variation on STANAG 5066 throughput is clearly evident.

5.6 Automatic Link Maintenance

Adaptive features are present throughout the 3G suite of protocols, including automatic channel selection prior to link establishment and code-combining ARQ for dynamic data rate adaptation during packet transmission. In the course of using an HF link, however, an increase in interference or pathloss may overwhelm the traffic protocol's ability to cope with the channel variations. In such a case, we need a mechanism to pause briefly while we find a suitable new frequency for the link, and then resume traffic with minimal disruption and without losing the connection. This is the role of the automatic link maintenance (ALM) protocol, which is used only on links set up using RLSU.

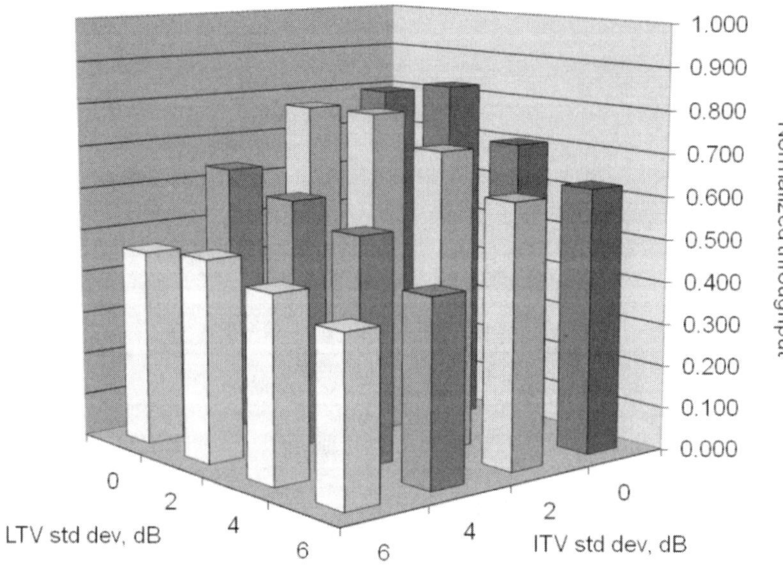

Figure 5.60 S5066/HDL+ throughput ratio, ITU-MLD, +20 dB mean SNR. (© 2007 IEEE. Reprinted with permission from [13].)

5.6.1 ALM PDUs

The protocol employs the robust BW1 waveform to convey the 48-bit PDUs shown in Figure 5.61. The fields in these PDUs are as follows:

- The first three bits are set to "1 1 1" to indicate the ALM protocol.
- The *countdown field* contains the number of times the PDU will be re-sent before the indicated change is to take effect. The sequence of LM PDUs is sent contiguously, ending with the PDU that contains a *countdown value* of 0. The number of repetitions of the LM PDU is chosen to reduce the probability that it will be missed by the other station(s) to an acceptable level.
- The *link ID field* in the PDUs is a hash of the station addresses engaged in the protocol, and is identical to that used in setting up the link (see Section 5.3.5.1).
- The *channel field* specifies one of the channels in the current set.

5.6.2 ALM Protocol Operations

ALM includes a family of functions that use these PDUs. Directed return to link setup (using the LM_Relink PDU) is mandatory for all implementations of ALM. The following optional functions are also defined:

- Probing of candidate alternative frequencies during a pause in traffic;
- Coordinated departure to suitable alternate frequencies;

LM_Probe

3	1	5	5	6	5	7	16
1 1 1	0	0 0 0 0 0	0 0 0 0 0	Link ID	0 0 0 0 0	Channel	CRC

LM_Request

3	1	5	5	6	5	7	16
1 1 1	0	0 0 0 0 0	0 0 0 0 1	Link ID	0 0 0 0 0	Channel	CRC

LM_Simplex

3	1	5	5	6	5	7	16
1 1 1	0	0 0 0 1 0	Countdown	Link ID	0 0 0 0 0	Channel	CRC

LM_Duplex

3	1	5	5	6	5	7	16
1 1 1	0	0 0 0 1 1	Countdown	Link ID	0 0 0 0 0	Channel	CRC

LM_WF_Ch

3	1	5	5	6	5	1	6	16
1 1 1	0	0 0 1 0 1	Countdown	Link ID	0 0 0 0 0	0	WF Code	CRC

LM_Relink

3	1	5	5	6	5	7	16
1 1 1	0	1 1 1 1 1	Countdown	Link ID	0 0 0 0 0	1 1 1 1 1 1 1	CRC

LM_Snap

3	1	21	7	16
1 1 1	1	First 21 HDL Acks	Channel	CRC

Figure 5.61 ALM PDUs.

- Negotiation of frequencies for duplex independent mode (i.e., different frequencies for transmitting and receiving);
- Renegotiation of waveform, data rate, and interleaver.

5.6.2.1 Relink

Either station in a PTP link (or the calling station in a PTM link) may initiate a return to link setup by sending LM_Relink PDU(s). All stations in the logical link will immediately return to scan. The station that originally set up the logical link will then initiate LSU to reestablish it. No response is made to this PDU; it is simply passed to the connection manager process at each receiving station.

5.6.2.2 Duplex Link Negotiation

At any time after a link is established on a traffic channel, including the time usually used for traffic setup, a station may send a sequence of LM_Duplex PDUs that indicates a channel on which other station(s) are requested to send future transmissions to the requesting station. All of the LM_Duplex PDUs are identical except for the countdown field. The sending station will continue to transmit on its current traffic channel after the change.

5.6.2.3 Probing of Candidate Traffic Channels

Two mechanisms are provided for the probing of candidate traffic channels: snap probing and extended probing. A snap probe may be extended at the discretion of the responding station, as described below.

Snap Probing Protocol

A *snap probe* measures a candidate traffic channel with minimum disruption to an in-progress HDL packet transfer. It is initiated by a station that is receiving an ARQ datagram by returning an LM_Snap PDU instead of an ACK PDU. The LM_Snap PDU contains 21 bits that replace up to 21 ACK bits from the expected ACK PDU, along with a channel number. The station that receives the LM_Snap PDU (the responding station) is expected to tune to the indicated channel and send an LM_Probe PDU. The other station (the requesting station) likewise tunes to the indicated channel and measures the signal quality of the Probe PDU. After the slot in which the probe is sent, both stations return to the original traffic channel, where the requesting station will do one of the following:

- Initiate a change to the recently measured channel;
- Initiate extended probing;
- Return the ACK that was preempted by the LM_Snap PDU, thereby resuming the interrupted ARQ protocol.

The responding station may send an LM_Request in place of the LM_Probe, which initiates extended probing.

Extended Probing Protocol

Extended probing is an open-ended handshaking protocol that can quickly evaluate several candidate traffic channels. It is initiated by either station in a point-to-point link by sending an LM_Request PDU that specifies a traffic channel to be evaluated. Any traffic protocol in progress is suspended for the duration of the extended probing. The extended probing is terminated by negotiation of the traffic channels for future transmissions.

Coordinated Departure to New Traffic Channels

Coordinated departure to new traffic channel(s) employs the LM_Simplex and/or LM_Duplex PDUs as appropriate to indicate a new frequency on which the sending station will listen for traffic. LM_Duplex PDU(s) indicate that the sending station will continue to send on its current transmit frequency until another frequency is negotiated. LM_Simplex PDU(s) indicate that the sending station will change its transmit frequency, as well as its receive frequency, to that indicated in the LM_Simplex PDU.

If the sending station detects protocol failure via timeout or other means after changing its receive frequency, it will execute the relink protocol, drop its logical link, and recommence link setup.

Negotiation of Waveform

The LM_WF_Ch PDU is used to negotiate waveform(s) to be employed on a logical link. The Waveform Code field values are identical to those used in TM_Confirm PDUs (see Table 5.5).

5.7 3G Multicasting

Multicasting is a technique for delivering traffic efficiently to a subset of network members. It falls between point-to-point techniques and broadcasting, and presents unique challenges in wired networks, line-of-sight wireless networks, and HF networks. Far from being a mere curiosity, multicasting is now the basis of a variety of popular applications, ranging from webcasts to the dissemination of situational awareness updates in tactical networks.

A multicast protocol for 3G networks had not been standardized at the time this book went to press, but the technology development was sufficiently mature enough to discuss here.

5.7.1 Introduction

When multicasting is offered at any layer of the protocol stack, it requires support from all lower layers. At the physical layer, multicasting requires either a broadcast channel or multiple point-to-point links. At higher layers, we are concerned with efficiently addressing the multicast destination stations; routing traffic so that redundant transmissions of packets are minimized while maximizing the probability that all destinations receive all packets; and collecting acknowledgments (when acknowledgments are required).

Multicast addressing schemes fall into two categories analogous to the net call, and group call addressing in 2G ALE. In many cases, a single collective address is assigned for a multicast. This is the approach taken in the Internet protocols (both version 4 and version 6), as well as 3G ALE. The alternative is to list explicitly the addresses of stations that are to receive the traffic.

Multicast routing in wired networks is concerned with forming a tree (for efficiency) or a mesh (for reliability) of links that connect the multicast source(s) to all destinations. Each router in the wired network then implements the computed topology by relaying incoming multicast packets via the correct subset of its outgoing links. By contrast, multicast routing in a (line-of-sight) wireless network requires determining which wireless nodes must rebroadcast multicast packets to ensure that all desired destinations can receive them. With the potentially global range of HF skywave links, rebroadcasting may not be required at all in HF multicasting, especially if the multicast source is able to send on multiple frequencies to reach both nearby and distant stations. Thus, the routing computation for HF multicasting may require only determining the list of frequencies on which a multicast is to be sent.

Finally, when a reliable multicast is required, we must provide a mechanism for multicast destinations to confirm receipt of messages, or to request retransmission of packets lost or received with uncorrectable errors. This becomes complicated when some stations must remain in radio silence (EMCON) for extended periods. This common requirement of tactical military networks (see, for example, STANAG 4406 [16]) is addressed by the P_MUL protocol.

5.7.2 P_MUL

P_MUL is an application-layer reliable multicast protocol developed for use by military and similar users of wireless networks. P_MUL, standardized in Allied Communications Publication 142 [17], was developed specifically to address networking applications that have low bandwidth and delayed acknowledgments (e.g., stations in EMCON).

As an application-layer protocol, P_MUL uses lower-layer protocols to transmit its PDUs over a multicast network. Since nodes under EMCON are not allowed to acknowledge messages, they are unable to use a reliable transport protocol, like TCP, for the transmission of messages. Therefore, P_MUL is based on the use of a connectionless transport protocol, such as UDP.

Although P_MUL is based on a connectionless transport protocol, it provides users with reliable connection-oriented multicast services. It enables the receivers to receive messages while being under EMCON restrictions. It ensures that the transmitter is informed about the timely completeness of the transmission of the messages after the receivers leave the EMCON status and, if required, enables the retransmission of any messages that were not properly received.

It is envisaged that P_MUL will be deployed in networks ranging in size from a few nodes to hundreds of nodes.

5.7.2.1 P_MUL PDUs

P_MUL uses four different PDUs for the data transfer as follows:

- Address_PDU identifies message addressees;
- Data_PDU carries message fragments;
- Ack_PDU acknowledges message reception;
- Discard_Message_PDU terminates the transmission of a specific message.

Address_PDU

The P_MUL transmitter generates an Address_PDU to announce intended recipients of a message and provides the Message_ID. This PDU and the Ack_PDU effect flow control of P_MUL packets. As P_MUL has a bounded PDU size and an unbounded the number of Destination_Entries, it is possible to split the complete addressing information into more than one Address_PDU using a MAP field in the PDU.

Data_PDU

The P_MUL transmitter generates the Data_PDU to carry each message fragment. This PDU includes the unique identifier of the message, the position of this Data_PDU within the ordered set of all Data_PDUs along with a fragment of the total message.

Ack_PDU

This PDU is generated by a receiving node identified to inform the transmitting node of the status of one or more messages received. It carries one or more Ack_Info_Entries. Each of these contains a message identifier (Source_ID and Message_ID) and a list of Missing_Data_PDU_Seq_Numbers (a list of those Data_PDUs not yet received). If this list is empty, the message identified by Source_ID and Message_ID has been correctly received.

Discard_PDU

The Discard_PDU is generated by a P_MUL transmitter to inform the receiving nodes that the transfer of a specific message has been terminated and no further PDUs of this message will be sent. Such situations can arise in the event of hardware error or message obsolescence. PDUs already received are to be discarded by the receiving node.

5.7.2.2 P_MUL Protocol Operation

A node initiates the transmission of a message by sending an Address_PDU containing a list of all the nodes that are to receive the message. The transmitting node must be informed by a management function about the operating mode of each receiving node (i.e., which receiving nodes are in EMCON). Based on this information, the transmitting node creates a list of non-EMCON receiving nodes, from which it will expect to receive Ack_PDUs.

After sending the Address_PDU, the transmitting node will send the Data_PDU(s). After sending the last Data_PDU of a message, the transmitting node initializes a Transmitter Expiry_Time Timer. If one or more of the receiving nodes is in the EMCON mode, the transmitting node also initializes an EMCON retransmission counter (EMCON_RTC). The transmitting node then enters the non-EMCON or EMCON retransmission mode (as directed by the management function).

Transmitter Expiry_Time Timer

The Transmitter Expiry_Time Timer tracks the time remaining until the message is considered invalid. Its initial value is announced in the Address_PDU. If all the receiving nodes addressed in the Address_PDU acknowledge receipt of the complete message before this timer expires, the timer is stopped.

If one or more of the receiving nodes have not acknowledged the receipt of the complete message when this timer expires, the transmitting node will send a Discard_Message_PDU that carries the Message_ID field of the expired message.

EMCON Retransmission Counter

The EMCON retransmission counter (EMCON_RTC) indicates the maximum number of times the complete message may be retransmitted while in the EMCON retransmission mode.

Non-EMCON Retransmission Mode

The transmitting node enters this mode only if at least one of the receiving nodes is in the non-EMCON mode. In this mode the transmitting node starts an Ack retransmission timer and listens for Ack_PDUs from non-EMCON receiving nodes. When this timer expires, the transmitting node re-transmits all Data_PDUs that remain unacknowledged by any non-EMCON node, preceded by an updated Address_PDU listing those nodes that had missing Data_PDUs. If the complete message has not been acknowledged by all non-EMCON nodes, the Ack retransmission timer is restarted, but with a timeout increased by a backoff factor. This cycle repeats until all non-EMCON receiving nodes have acknowledged the complete message. If there are receiving nodes in the EMCON mode, the transmitting node then enters the EMCON retransmission mode.

When a transmitting node receives an Ack_PDU from a non-EMCON or an EMCON receiving node (after leaving EMCON) indicating that it has received the complete message, the transmitting acknowledges this Ack_PDU by sending a modified Address_PDU that omits the address of that receiving node. After all receiving nodes have acknowledged the whole message, an Address_PDU containing an empty list of Destination_Entries is sent.

EMCON Retransmission

The transmitting node may enter EMCON retransmission mode when any receiving nodes are in the EMCON mode. Upon entering this mode, the transmitting node initializes the EMCON retransmission timer (EMCON_RTI). Each time this timer expires, the transmitting node resends the Address_PDU and all of the Data_PDUs, increments the EMCON retransmission counter, and restarts the EMCON_RTI if the counter has not reached its maximum value.

As receiving nodes in EMCON leave that state, they send Ack_PDUs, which are noted by the transmitting node. When all EMCON nodes have responded with an Ack_PDU, the transmitting node will either enter non-EMCON retransmission mode if any of the receiving nodes have missing segments, or send an empty Address_PDU if all receiving nodes have acknowledged the whole message.

5.7.3 MDL

To support P_MUL multicasting, the 3G HF subnetwork must provide a one-to-many delivery service. 3G ALE provides a multicast calling mode, but the 3G data link protocols presented so far are for point-to-point applications only. In this section, we present a proposed multicast data link (MDL) protocol.

P_MUL requires only a best effort datagram service, with acknowledgments handled at the application layer. However, as we have seen with supporting TCP over HF networks, we may obtain better performance if the link layer also provides a retransmission mechanism appropriate for the HF channel. Therefore, we also discuss a 3G multicast protocol with embedded retransmissions, the multicast data link with NAKs (MDLN) protocol, for non-EMCON users.

5.7.3.1 MDL Protocol

The protocol stack for multicast applications over HF is shown in Figure 5.62. Here, we see that the MDL protocol is added alongside the point-to-point 3G data link protocols HDL and LDL. MDL shares many of the characteristics of the other 3G data link protocols:

- Robust burst waveforms;
- Code combining to reduce the number of retransmissions required;
- MDL burst length is specified in the TM or FTM traffic type field.

Unlike the 3G ARQ protocols, MDL links will employ the one-way PTM link setup protocol.

MDL PDUs

The BW2 and BW3 burst waveforms used in HDL and LDL offer a useful range of throughput and robustness. The proposed MDL uses these existing burst waveforms:

- MDL-5K is a high-speed mode that uses 24-packet BW2 transmissions;
- MDL-512 is a more robust mode that uses a stream of 512-byte BW3 bursts;
- MDL-32 is a very robust mode that uses a stream of 32-byte BW3 bursts.

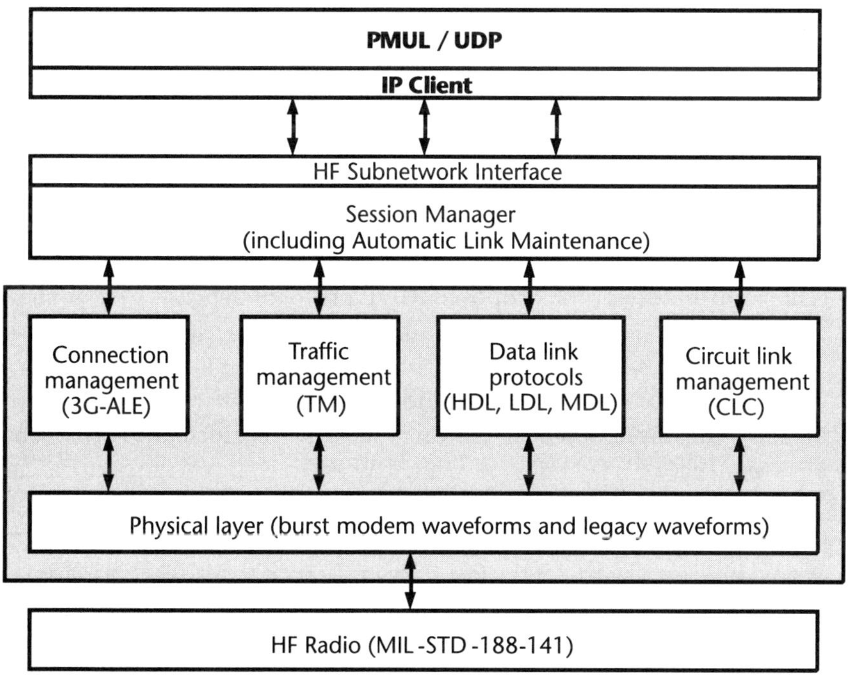

Figure 5.62 3G Protocol suite with multicast support.

In each case, the BW2 or BW3 bursts are created from payload data, as previously described. The separate sets of FEC output bits are produced for each packet of the message (four sets for BW2 and two for BW3). In MDL, unlike HDL or LDL, the entire Bitout_0 sequence (the first set of encoded bits for all packets of the message) is sent in a single, continuous transmission. After completing transmission of the first burst (BW2 or BW3), the next burst begins immediately, and so on until the Bitout_0 sequence for the entire message has been sent.

MDL Protocol Operation

When a message is to be sent using MDL, the session manager specifies the number of transmissions of that message to be sent, along with which MDL mode is to be used. The MDL protocol then sends the entire Bitout_0 sequence for that message as the first transmission. If more than one transmission was specified, the next transmission will contain the Bitout_1 sequence, and so on, cycling through the FEC-encoded message sequences the requisite number of times. When the specified number of transmissions has been sent, MDL reports completion and the session manager can then request that another message be sent, wait for ACKs, or direct the connection manager to drop the PTM link.

The receiver must determine which FEC-encoded sequence is arriving using the history of the multicast link as well as the contents of the incoming data. No control packets are sent to indicate the FEC phase. Just as for HDL and LDL, the receiver incrementally combines soft decisions from all received versions of each packet to attempt to recover error-free packets.

The transmission of the entire message in one code phase before retransmitting any packets in another code phase offers two benefits: (1) some time diversity, which should improve code combining performance, and (2) the ability to deliver the entire message to the client early if the message has been decoded error-free before all of the scheduled repetitions.

Each transmission begins with a TM, FTM, or FLSU PDU that indicates the MDL mode that will be used in the remainder of the transmission.

5.7.3.2 MDL Performance[7]

The performance of the proposed MDL protocol depends first of all on the robustness of the PDUs. Measurement and modeling of the selected BW2 and BW3 bursts yield the probability of packet error versus SNR curves in Figures 5.63 and 5.64. The notation used in the figures indicates the number of Bitout sequences that are sent. For example, BW2 ×1 indicates the error performance after sending only the first sequence of BW2 packets, while BW3-32 ×2 shows the performance after sending both sequences of BW3-32 packets.

We estimate the performance of the MDL protocol in three scenarios: (1) regional multicast within a theater of operations, (2) strategic dissemination of Emergency Action Messages, and (3) long-haul multicast of an Air Tasking Order, with acknowledgments.

7. This section updates an earlier study presented at MILCOM 2008 [18].

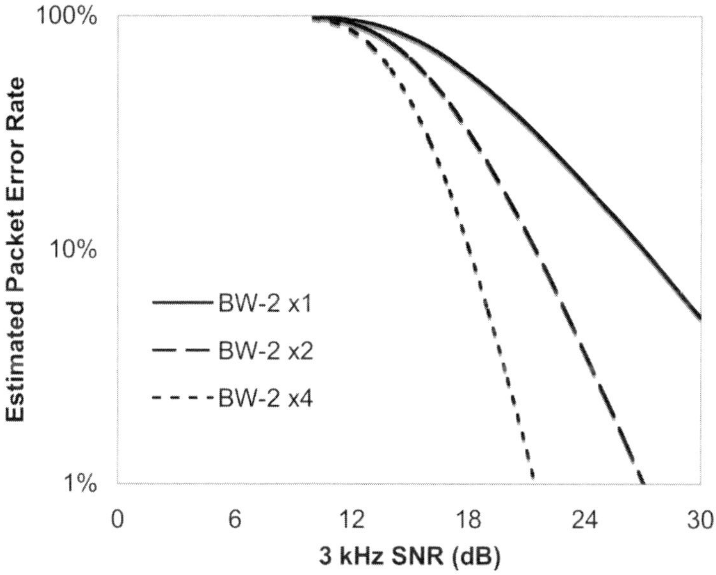

Figure 5.63 Packet error probability of MDL BW2 PDU.

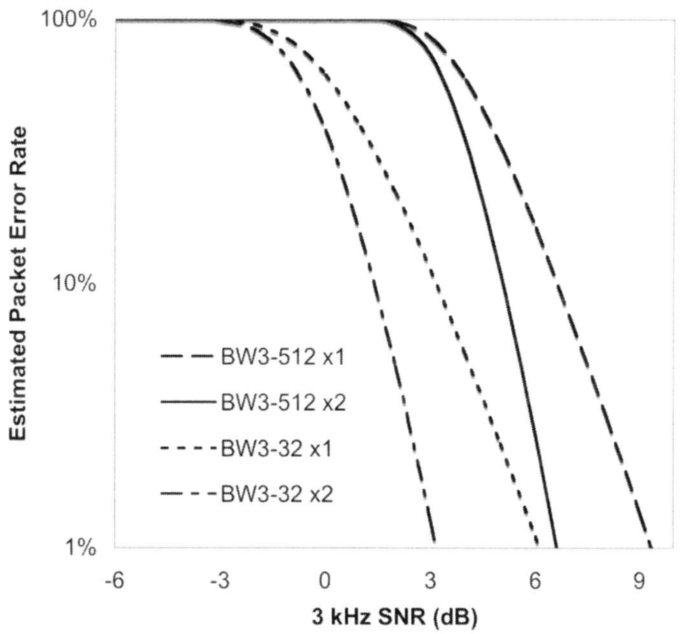

Figure 5.64 Packet error probability of MDL BW3 PDUs.

In each scenario, we employ the DoD-validated [7] NetSim approach to predict SNR to each receiver. For these simulations, we used the first-decile SNR (i.e., the SNR that will be exceeded 90% of the time). These are therefore quite conservative predictions compared to using median (fifth-decile) SNR values. For each hour of the day, the optimum frequency was used.

We evaluate each scenario for three ionospheric conditions: (1) summer (July) with low solar activity (SSN = 10), (2) autumn (October) with high solar activity (SSN = 130), and (3) spring (April) with moderate solar activity (SSN = 70).

Regional Multicast Scenario

In this scenario, we assume that an NVIS path is used to deliver situational awareness updates every five minutes from a ship standing offshore to a Marine regiment that is ashore. Each update is a compressed message, 10 kB in size. Six HF radios, distributed over the landing zone, receive the multicasts and forward the data into the tactical line-of-sight radio networks ashore. No acknowledgments are returned to the ship. Note that because this is a compressed message, all of the message packets must be received error-free before decompression can be successful; partial delivery is not possible.

This message is large enough to make good use of the highest-throughput waveform, MDL-5K. However, for SNR less than about 18 dB, MDL-512 provides better throughput [18]. For reliable message delivery after sending all four MDL-5K sequences once, we need SNR of at least 25 dB. A 1 kW shipboard transmitter provides SNRs ashore that range from 20 to 34 dB over the conditions simulated, which suggests that sometimes MDL-5K would not be reliable. Therefore, we also evaluate the use of MDL-512 for improved robustness.

Figure 5.65 shows the message delivery probabilities over 24 hours during especially challenging propagation conditions: July, which has low solar activity. It is clear that MDL-5K, even with four transmissions of the message, was not reliable during many hours of the day. However, MDL-512 was 100% reliable with only a single message transmission, in this and all other conditions simulated.

The higher reliability of MDL-512 comes at a cost in speed, however. A single transmission of the message using MDL-512 requires 150 s. For MDL-5K, a single

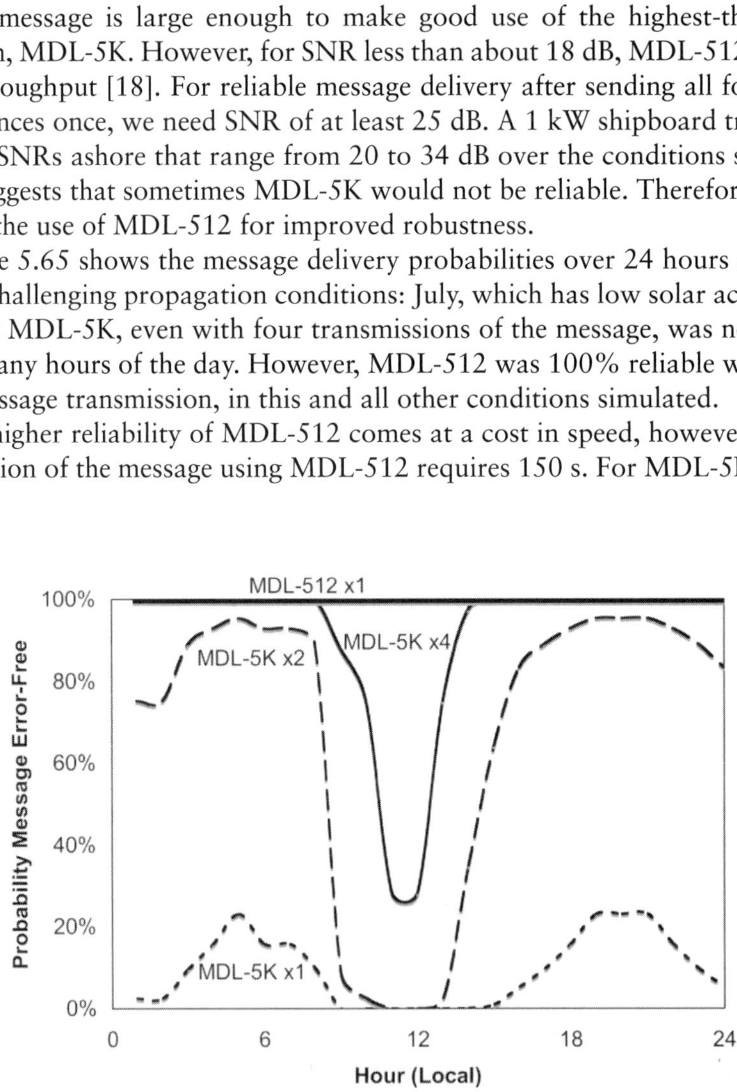

Figure 5.65 Message delivery probability in NVIS scenario, July, SSN = 10.

transmission requires only 22.5 s, although successful reception at this point in the process was rare; the full fourfold transmission requires 90 s.

EAM Scenario

In this "Dr. Strangelove" scenario, we have 24 aircraft (perhaps strategic bombers) distributed over Alaska, western Canada, and the western United States (Figure 5.66). Four high-power (4 kW) base stations simulcast a 32-byte emergency action message (EAM) to these aircraft on frequencies that span the HF spectrum. We can imagine that the EAM is ordering the bombers to return to base rather than start World War III, so it is critically important that the aircraft receive the EAM!

The size of the message fits nicely within the 32-byte payload of MDL-32, and we will employ this most robust mode of the MDL protocol to improve the chances of the message reaching the bombers. The message will be sent repeatedly to ensure that all of the aircraft eventually receive it. In accordance with the MDL-32 protocol, the two differently-coded sequences are sent in alternation.

Since every aircraft will eventually receive the message, message delivery probability is not an interesting performance metric. Instead, we evaluate the reliability of message delivery *within the first four repetitions of the message*. In the case of low solar activity during the summer, we find that every aircraft will receive the message within four repetitions (7.5 s) at every hour of the day. The more challenging condition is found at high solar activity in October. The *worst case* reliability for any single aircraft under these conditions (after four transmissions) is plotted in Figure 5.67. The two aircraft most distant from the transmitters do not receive the EAM after four transmissions in the predawn hour. They do, however, eventually receive the message. All of the other aircraft have 100% reliability with four repeti-

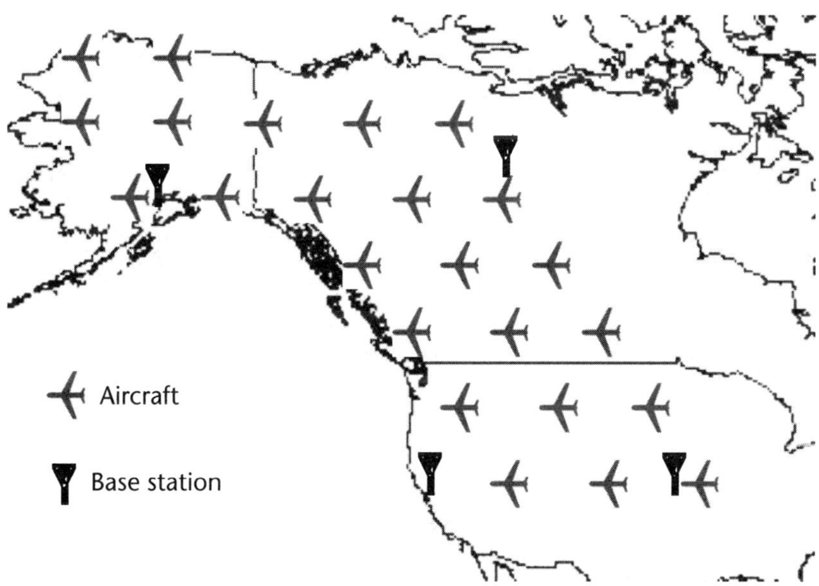

Figure 5.66 Geography of the EAM scenario.

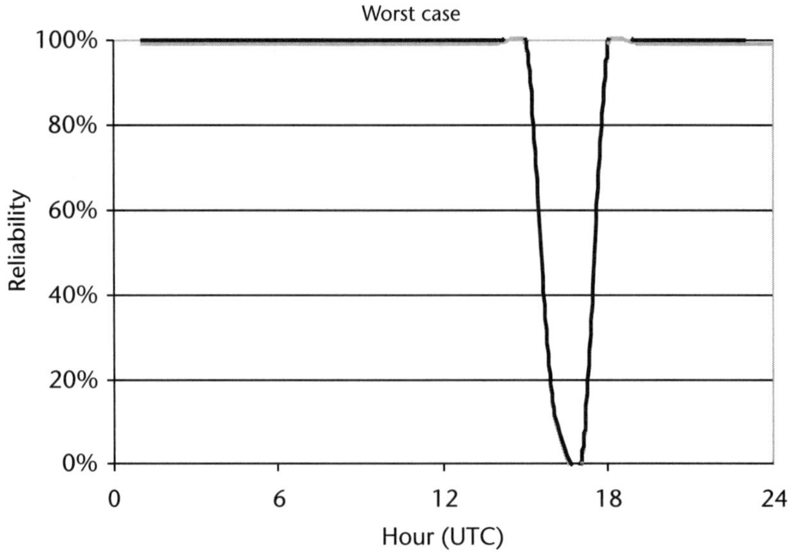

Figure 5.67 Reliability of EAM receipt after four transmissions (October, SSN 130).

tions at all hours of the day. (Springtime with moderate solar activity has similar performance.)

ATO Scenario

Our third scenario is also strategic in scope, but now includes acknowledgments. Here, we have a single base station that multicasts air tasking orders (ATOs) to eight aircraft distributed throughout a theater of operations. A new ATO is multicast every 6 hours. The entire ATO is repeated twice an hour until all of the recipients have acknowledged error-free receipt of the entire 100-kB message. MDL-5K is the preferred multicast mode due to the large size of the message. All four encodings can be sent in 14 minutes. MDL-512 is more robust, but requires 24 minutes for a single transmission of the ATO.

Acknowledgments are returned by the aircraft using FLSU to set up a point-to-point link to the base station and LDL to return the P_MUL ACK.

Under most conditions, MDL-5K ×4 provides 97% to 100% reliability in delivering the ATO on the first try. However, during the summer when solar activity is low, there is a dip in reliability around noon local time (Figure 5.68). MDL-512 provides full reliability under all conditions.

Perversely, the high reliability of delivering the message to all destinations results in congestion at the end of the multicast, as all of the aircraft attempt to establish links with the base station to return acknowledgments. However, the number of contending aircraft is small, and the congestion is quickly resolved. Figure 5.69 shows the cumulative percentage of ACKs returned, averaged over all hours, versus time elapsed from the end of the multicast. The two extreme conditions are shown: July with low solar activity and October with high solar activity.

In all cases, over 90% of the acknowledgments were returned within one minute, and 100% were returned within 3 minutes.

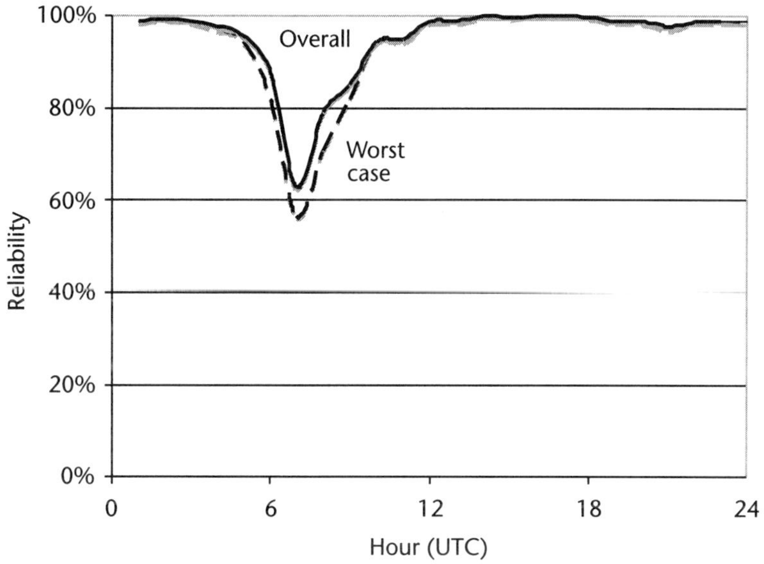

Figure 5.68 ATO reliability.

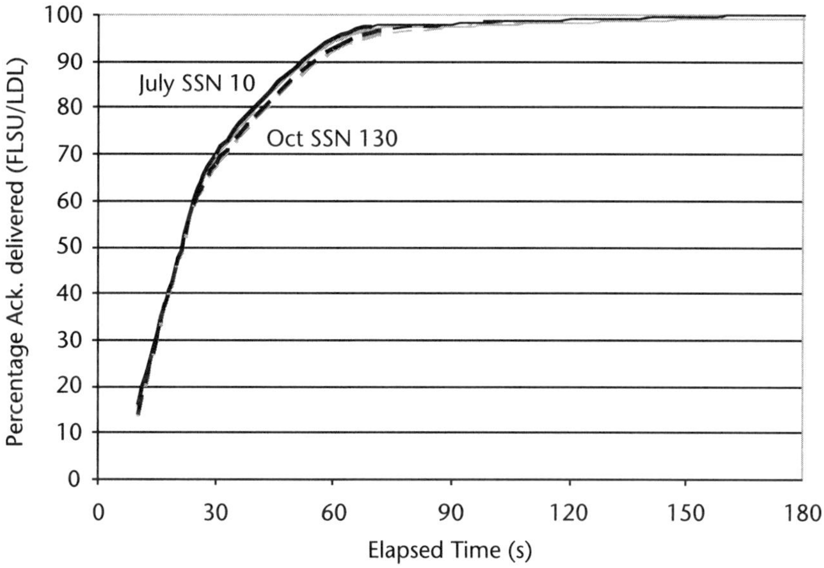

Figure 5.69 ATO acknowledgment time.

5.7.4 MDLN Protocol

The MDL protocol employs the incremental redundancy common to 3G data link protocols in an open loop fashion: the entire message is sent using the Bitout_0 encoding, optionally followed by transmission(s) using the other encoding(s) of the message. A station that decodes the entire message error-free after the first transmission can deliver the message and drop out of the multicast after that first transmission. Otherwise it must process the retransmissions and attempt to correctly decode the message after combining additional received information.

The multicast data link with NAKs (MDLN) brings the closed loop form of incremental redundancy to bear. In MDLN, each forward transmission is followed by a pause, during which receivers that were not able to decode that transmission emit a very robust pseudonoise (PN) PSK symbol sequence to request retransmission (Figure 5.70).

All receivers share the NAK slot. (Detection of the PN NAK sequence is sufficiently robust to allow any number of NAKs to overlap during the slot.) When the sender detects an NAK, it sends additional redundancy bits. Thus MDLN, like the point-to-point ARQ protocols, sends only enough redundancy to convey the message error-free. Additional details of the MDLN conceptual design may be found in [19].

5.7.5 Conclusion

The proposed MDL/MDLN protocol family promises to provide robust multicasting in 3G HF networks in both tactical and strategic applications. The MDL-32, MDL-512, and MDL-5K options offer a wide range of speed and robustness. However, in the scenarios examined here, the MDL-5K was sometimes too unreliable and had to be backstopped by MDL-512. MDLN augments the robust forward error correction capabilities with link layer acknowledgments. Discussions are ongoing in the standards community regarding which modes of 3G multicast will be adopted.

5.8 3G Performance in Internet Applications

As the modern world relies increasingly on Internet access, it is important to understand the characteristics of Internet applications and protocols, and to examine how they interact with and can be supported by 3G HF technologies.

5.8.1 Characteristics of Internet Applications

The protocols that support common Internet applications may be considered in terms of their protocol layers. Above the subnetwork interface to the HF radio network we find (in ascending order) the Internet layer, the transport layer, and the application layer.

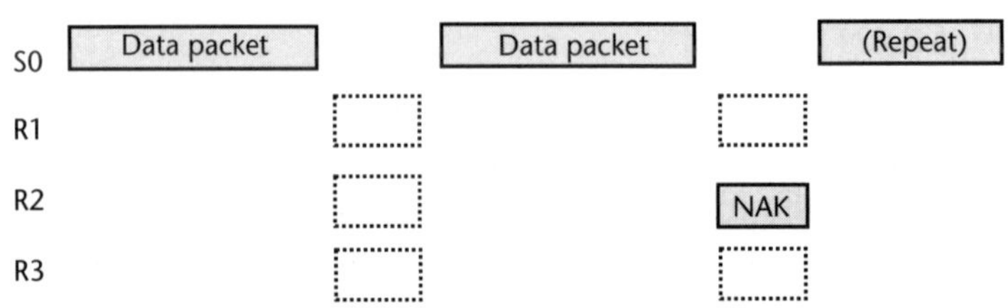

Figure 5.70 MDLN NAK operation [19].

- The protocols at the *Internet layer*, including the Internet protocol (IP) as well as the Internet control message protocol (ICMP) and Internet group management protocol (IGMP), generally require only best effort service from the HF subnetwork, and consequently do not interact strongly with the HF subnetwork protocols.
- At the *transport layer*, the user datagram protocol (UDP) is likewise easy to support over the HF subnetwork, but the transmission control protocol (TCP) tends to interact strongly and poorly with the HF subnetwork service.
- The *application layer* protocols support the familiar Internet applications, such as sending and retrieving email, downloading web pages, streaming voice and video, and so on. Most applications are interactive, in the sense that the user expects a prompt response to his actions. Email differs in user expectations: we recognize that it sometimes takes minutes or even hours for an email message to be delivered through the Internet, although messages are most often delivered in seconds. This user acceptance of occasional delays in email delivery made email a natural fit for the occasional outages experienced in an HF subnetwork service. Thus, email has been described as the "killer app" that popularized HF as a modern data network.

5.8.1.1 Application Protocol Characteristics

Unlike the lower-layer protocols, Internet application-layer protocols mostly employ plain text PDUs. For example, Figure 5.71 shows the SMTP handshake involved as a client computer myhost.mycollege.edu submits a message from myname@mycollege.edu to the server server.agency.gov.

The important characteristics of most application protocols are visible here:

- Most of the application-layer PDUs are relatively short (e.g., "220 server.agency.gov ESMTP Sendmail ready"). In this example, only the email message itself (abbreviated here to "large email message goes here") is of appreciable size. Such short transmissions are comparable in size to the TCP and IP headers (around 40 bytes total), so their efficiency is 50% or less, even before

Server	**Client**
220 server.agency.gov ESMTP Sendmail ready	
	HELO myhost.mycollege.edu
250 server.agency.gov	
	MAIL From:<myname@mycollege.edu>
250 <myname@mycollege.edu>... Sender ok	
	RCPT To:<recipient@somewhere.gov>
250 <recipient@somewhere.gov>... Recipient ok	
	DATA
354 Enter mail, end with "." on a line by itself	
	<large email message goes here> .
250 Message accepted for delivery	
	QUIT
221 server.agency.gov closing connection	

Figure 5.71 SMTP example.

considering interactions with the HF subnetwork. Both 2G and 3G protocols are most efficient when conveying large payloads.

- Because of the short messages, link turnarounds at the application layer are frequent. This, of course, necessitates frequent link turnarounds down through the stack of protocols, and we know from Chapter 3 that link turnarounds can be costly in the HF subnetwork, especially when using 2G technology. 3G protocols are more nimble in reversing the direction of the physical link, but must renegotiate the direction of data flow on the logical link each time the application layer reverses direction.

Recognizing the impact that these application characteristics have on performance through an HF subnetwork, HF-friendly versions of several popular protocols have been developed and standardized. For example, HMTP is the HF-friendly version of SMTP. It mandates use of the ESMTP command pipelining extension, wherein all of the commands from the client are sent in a single burst (as shown in Figure 5.72). By breaking the lockstep exchange of short transmissions, we address both of the concerns noted above.

5.8.1.2 TCP Characteristics

TCP provides an end-to-end reliable connection through the variety of subnetworks encountered along a path through the Internet. The common versions of TCP implement a go-back-N ARQ protocol with an adaptive timeout mechanism. Both of these present problems in using TCP over HF subnetworks.

- The go-back-N ARQ protocol retransmits all TCP segments following a lost segment. If we are using a link-layer ARQ protocol in the HF subnetwork, we should never lose a TCP segment, but poor channel quality can require many link-layer retransmissions for successful delivery. Once the link-layer delay exceeds the TCP timeout, TCP will resubmit the lost segment—and eventually all those following it—for delivery over the struggling HF link. This, of course, exacerbates the problem, adding congestion delay to the link-layer delay.
- The adaptive timeout mechanism in TCP gradually learns the round-trip time (RTT) through the path to the destination, using a smoothed average of RTT

Server	Client
220 server.agency.gov ESMTP Sendmail ready	
	EHLO myhost.mycollege.edu MAIL From:<myname@mycollege.edu> RCPT To:<recipient@somewhere.gov> DATA <large email message goes here> . QUIT
250 server.agency.gov 250 <myname@mycollege.edu>... Sender ok 250 <recipient@somewhere.gov>... Recipient ok 354 Enter mail, end with "." on a line by itself 250 Message accepted for delivery 221 server.agency.gov closing connection	

Figure 5.72 HMTP example.

measurements. The starting RTT estimate was historically 3 s, but that was reduced to 1 s by RFC-6298 [20] in 2011. Whenever the retransmission timer expires, the timeout value is doubled and the unacknowledged segment is retransmitted. The TCP timeout is set to the RTT estimate plus four times the measured variation in RTT. As we shall see, the initial RTT estimate is too low for HF subnetworks, and the variation term can become quite large.

5.8.2 Interaction of Internet Protocols with HF Data Links

When a TCP connection is first attempted through an HF subnetwork, we will probably encounter a delay of 10 s or more while ALE establishes the HF link (see Section 5.3.6.2). During this interval an RFC-6298-compliant TCP will time out at 1s, 3 s, and 7 s, (and perhaps 15 s), resending the SYN packet each time. Once the connection through the HF subnetwork is running, the round-trip time estimate should gradually settle to a steady-state value. RFC-6298 allows a maximum value for the RTT estimate, but requires that this value be at least 60 s.

The dynamic nature of the HF channel will be reflected in large variations in RTT as measured by TCP. For example, the cycle time of the STANAG 5066 protocol ranges up to 120 s, so when the HF data link has to retransmit a single lost frame, we will see quite large variation in RTT measurements.

5.8.2.1 Simulation Study of HF Support for Internet Applications

Simulation studies have evaluated the magnitude of these interactions between Internet and HF protocols [21, 22]. For this work, the STANAG 5066 ARQ, HDL, LDL, TCP, SMTP, and HMTP protocols were implemented within the NetSim simulation framework. (The application-layer protocols were script-driven, not fully implemented.) Detailed discussions of the interactions may be found in the conference papers [21, 22]. Here, we will take a different perspective, examining the overall impact of passing unmodified Internet traffic through an HF subnetwork, versus terminating the Internet protocols at an HF gateway and employing HF-friendly protocols within the HF subnetwork.

The simulated channel for this study was similar to that described in Chapter 2, with lognormal-distributed fluctuations about the average SNR. Here we simulated 4 dB standard deviation for these fluctuations, with a 10-second autocorrelation time constant. For each simulation, the average SNR is held constant; we do not simulate the diurnal change. Link establishment is also not simulated; in effect, the link was established before the simulation began.

We focus here on email performance in 2G and 3G networks. A steady load of 5,000-byte messages is presented to the application, and we measure the number of messages delivered per hour. Two client protocol stacks are considered: SMTP with TCP represents unmodified Internet traffic, and HMTP with no transport protocol is typical of HF-friendly traffic.

In Figure 5.73 we see that the 2G subnetwork (using STANAG 5066 ARQ and the MIL-STD-188-110B modem) achieves about three times higher throughput for email delivery when an HF-friendly protocol is used than when the usual Internet protocols are used. Also note the sharp rise in throughput when the SNR reaches

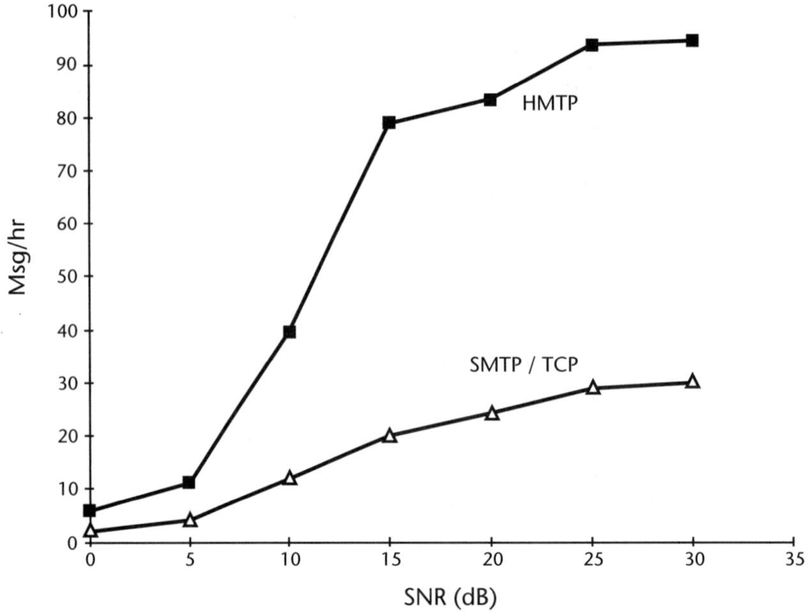

Figure 5.73 HMTP versus SMTP/TCP over STANAG 5066.

10 dB or better (a voice-quality link). Of course, the highest throughput is achieved when the full 9600 bps capability of the modem can be used.

We see a similar performance difference in the 3G network (Figure 5.74). As expected, the 3G network offers much better performance than 2G at low SNR, and smoothly increases throughput as SNR increases. High-SNR performance is limited by the 4800-bps data rate of the HDL waveform. HDL+ was not evaluated, but should offer a substantial throughput improvement over HDL.

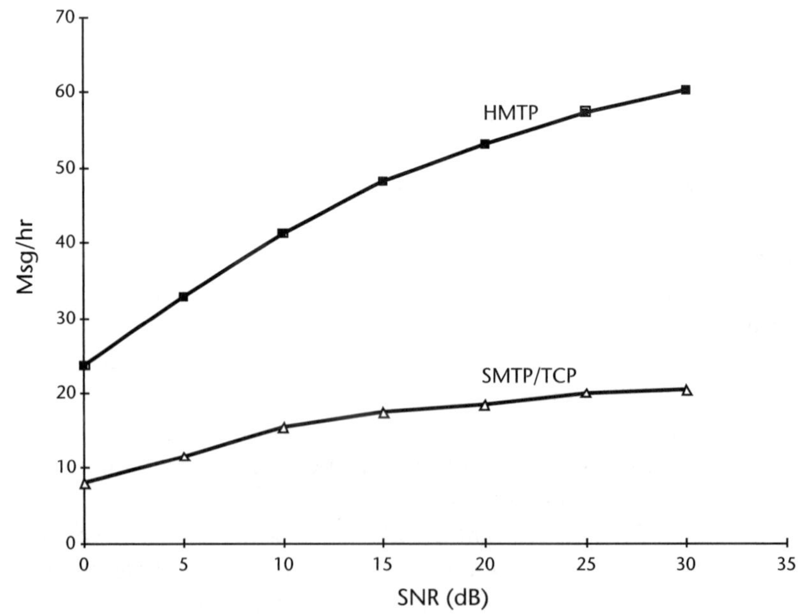

Figure 5.74 HMTP versus SMTP/TCP over HDL.

5.8.2.2 Measurements of Internet Application Performance in 3G HF Radio Networks

Koski et al. [23] report laboratory measurements of Internet traffic delivery by 3G HF radio networks. These experiments used synthetic IP test applications intended to "simulate the traffic profiles generated by different [military] C3I applications."

UDP Exchange of Situational Awareness Packets

A 10-station mesh network was formed by tying all of the radio RF ports together through attenuators.

- Each radio includes an internal channel simulator, all of which were programmed identically for an AWGN channel with the SNR indicated in the graph;
- The frequency pool included four channels;
- The PC controlling the experiment generated 150-byte packets at a selected mean rate, with uniformly distributed interarrival times from 0 to twice the mean interarrival time;
- Each packet was issued to a randomly chosen radio and destined for another randomly chosen radio;
- Links were established using FLSU and the packet was delivered using either xDL (LDL or HDL as selected by the radio) or HDL+.

Figure 5.75 shows the latency for delivering the UDP packet as a function of offered load, channel SNR, and protocol used. Note the improved performance of HDL+ in a high-SNR channel.

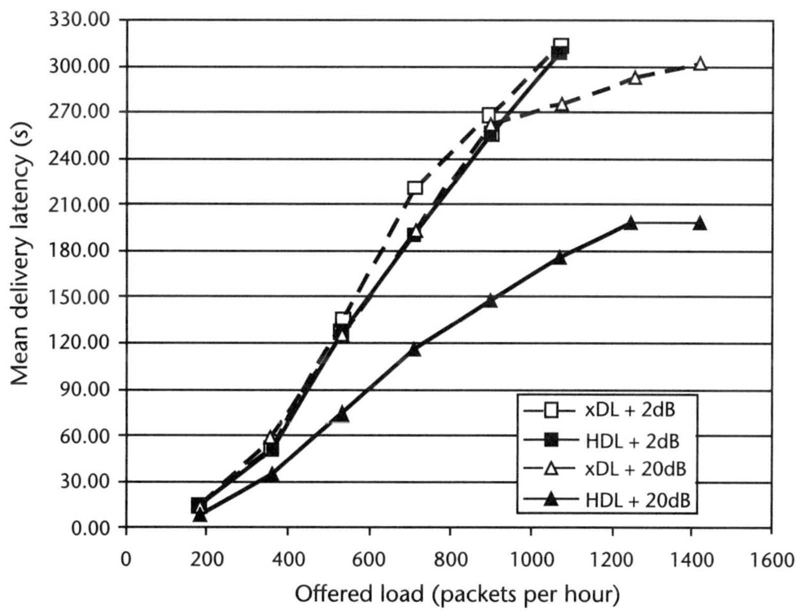

Figure 5.75 Measured latency for UDP packet delivery in a 10-station 3G HF network. (After [23].)

Point-to-Point TCP Throughput

Throughput on a single TCP connection was measured for a range of message sizes, from 1 through 1 million bytes. For comparison with earlier results, we show here the throughput measured for 5,000 byte messages. Again, an AWGN channel is used, and throughput is measured after the link and the TCP connection are established. In Figure 5.76, we see that the measured performance for SNR up to +5 dB is consistent with the simulated performance of SMTP/TCP over HDL (Figure 5.74). Above +5 dB, HDL+ provides a substantial boost in throughput.

Point-to-Point FTP Throughput

Finally, application throughput was measured for the following sequence:

- An FTP client logged into an FTP server via a 3G HF radio link.
- The client uploaded a file to the server, then downloaded a file of the same size.
- The client logged off from the server.

Throughput in bps is shown in Figure 5.77 as a function of file size and channel SNR.

5.9 Field Testing

3G HF technology has been proven in the field for more than a decade. When the 10th Mountain Division deployed to Afghanistan in support of Operation Enduring

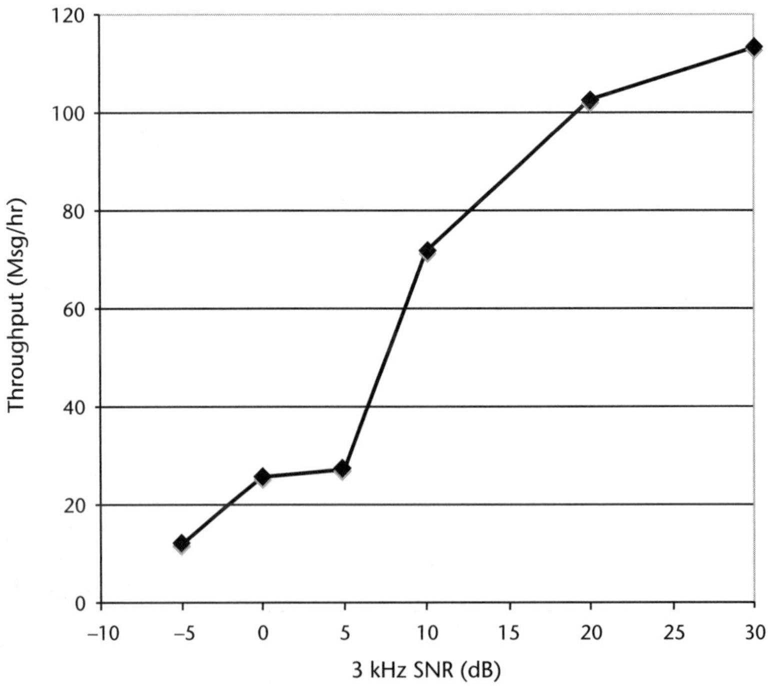

Figure 5.76 Measured TCP throughput of 5,000-byte messages, point-to-point 3G link. (After [23].)

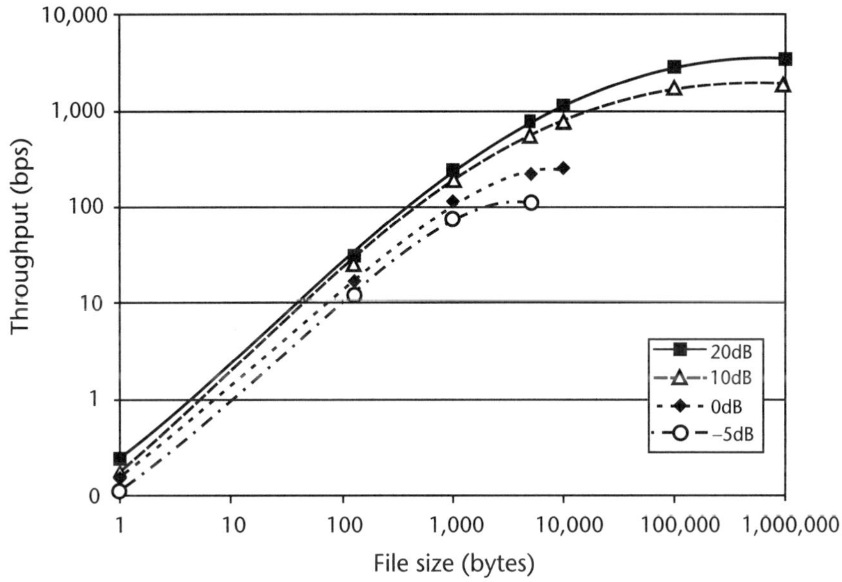

Figure 5.77 Measured FTP throughput over a point-to-point 3G link. (After [23].)

Freedom in late 2001, they were delighted with the capabilities of their 3G HF radios (from first-hand reports to the authors upon their return).

Field testing of 3G HF radios by Harris field engineers has also yielded positive user observations (reported in [24]):

- Communications planning for a 3G HF network is somewhat simpler than for a 2G network, with fewer parameters to set.
- The speed and responsiveness of basic operations are greatly improved when compared to 2G.
- Internet-style instant messaging is comfortable and effective with the 3G technology.
- Communications benefit greatly from the increased robustness of the STANAG 4538 protocols. Text messaging suffered little or no noticeable delay, even in SNRs down to 0 dB.

5.10 Summary of 3G HF Technology

The goals of 3G HF technology were to link faster and at lower SNR, to carry more traffic, and to support larger networks than 1990s HF technologies offered. These goals were achieved. 3G HF radios are now widely fielded, finding application most often in tactical networks where the ability to pass data at lower SNR offers longer battery life and a reduced RF footprint. Work continues on multicasting and other extensions, but the core of 3G HF technology is now well-established and field proven.

References

[1] Standardization Agreement 4538, *Technical Standards for an Automatic Radio Control System for HF Communication Links*, NATO, 2007.

[2] Chamberlain, M., W. Furman, and E. Leiby, "A Scaleable Burst HF Modem" *Proceedings of HF98, The Nordic Shortwave Conference*, Fårö, Sweden, 1998.

[3] Furman, W. N., "Robust Low Bit Rate HF Data Modems," *Proceedings from the IEE Seventh International Conference On HF Radio Systems and Techniques*, University Of Nottingham, U.K., July 1997.

[4] Ma, H. H., and J. K. Wolf, "On Tail-Biting Convolutional Codes," *IEEE Trans. Commun.*, Vol. COM-34, February 1986, pp. 104–111.

[5] Recommendation ITU-R F.1487, Testing of HF Modems with Bandwidths of Up to about 12 kHz Using Ionospheric Channel Simulators.

[6] Johnson, E. E., "Fast Propagation Predictions for HF Network Simulations," *Proceedings of MILCOM '97, IEEE*, Monterey, CA, 1997.

[7] Bullock, R. K., "SCOPE Command High Frequency (HF) Network Simulation Model (NetSim-SC) Verification and Validation Report," Defense Information Systems Agency, Joint Interoperability Test Command, Ft. Huachuca, AZ, January 1998.

[8] Johnson, E. E., "Simulation Results for Third-Generation HF Automatic Link Establishment," *Proceedings of MILCOM '99, IEEE*, Atlantic City, NJ, 1999.

[9] Johnson, E., T. Kenney, M. Chamberlain, W. Furman, E. Koski, et al., "U.S. MIL-STD-188-141B Appendix C: A Unified 3rd Generation HF Messaging Protocol," *Proceedings of HF98, The Nordic Shortwave Conference*, Fårö, Sweden, 1998.

[10] Wicker, S. B., *Error Control Systems for Digital Communication and Storage*, Upper Saddle River, NJ: Prentice-Hall, 1995.

[11] Chamberlain, M. W., and W. N. Furman, "HF Data Link Protocol Enhancements Based on STANAG 4538 and STANAG 4539, Providing Greater Than 10KBPS Throughput Over 3kHz Channels," *Proceedings of the Ninth International Conference on HF Radio Systems and Techniques*, IEE Conference Publication #493, University of Bath, U.K., June 2003.

[12] Furman, W., and J. Nieto, "Understanding HF Channel Simulator Requirements in Order to Reduce HF Modem Performance Measurement Variability," *Proceedings of the 2001 Nordic Shortwave Conference (HF 01)*, Fårö, Sweden, 2001.

[13] Batts, W. M., W.N. Furman, and E. Koski, "Channel Quality Variation as a Design Consideration for Wireless Data Link Protocols," *Proceedings of IEEE Military Communications Conference (MILCOM) 2007, IEEE*, Orlando, FL, October 2007.

[14] Watterson, C. C., J. R. Juroshek, and W. D. Bensema, "Experimental Confirmation of an HF Channel Model," *IEEE Transactions on Communication Technology*, Vol. COM-18, No. 6, December 1970.

[15] FED-STD-1052, "Telecommunications: HF Data Modems," General Services Administration, August 7, 1996.

[16] Standardization Agreement 4406, *Military Message Handling System*, NATO, 2006.

[17] Allied Communication Publication 142, P-MUL—A Protocol for Reliable Multicast Messaging in Bandwidth Constrained and Delayed Acknowledgement (EMCON) Environments, February 2001.

[18] Johnson, E. E., "IP Multicasting in HF Radio Networks," *Proceedings of MILCOM 2008*, San Diego, CA, 2008.

[19] Koski, E., "Concepts for a Reliable Multicast Data Link Protocol for HF Radio Communications," *Proceedings of MILCOM 2005*, Atlantic City, N.J., 2005.

[20] Paxson, V., et al., "Computing TCP's Retransmission Timer," RFC-6298, June 2011.

[21] Johnson, E. E., "Interactions Among Ionospheric Propagation, HF Modems, and Data Protocols," *Proceedings of the 2002 Ionospheric Effects Symposium*, Alexandria, VA, 2002.

[22] Johnson, E. E., "Interoperability and Performance Issues in HF E-Mail," *Proceedings of MILCOM 2001*, McClean, VA, 2001.

[23] Koski, E., W. Batts, Jr., and T. Benedett, "Effective Communications for C3I Applications Using Third-Generation HF," *Proceedings of the Nordic Shortwave Radio Conference (HF '04)*, Fårö, Sweden, 2004.

[24] Koski, E., et al., "STANAG 4538 Implementation and Field Testing Lessons Learned," *Ninth International Conference on HF Radio Systems and Techniques*, University of Bath, UK, June 23–26, 2003.

CHAPTER 6

Wideband HF

Years ago, it was believed that the use of HF radio would decline as competing satellite communications systems were brought online, but the recurring cost of these competing systems—in combination with the lower nonrecurring cost, the improved reliability, and the increased data rates achievable over 3-kHz HF channels—brought HF back into the forefront of long-range wireless communications. In recent times, the challenge for HF has been to provide data rates high enough to support services that users now consider to be essential. Instead of facing obsolescence, the question now facing the HF community is "How can we achieve even higher data rates over HF?"

6.1 Introduction

For many years, both voice and data communications in the HF radio bands have usually been restricted to channel bandwidths no wider than 3 kHz (although occasionally a single transmitter was allowed to occupy two adjacent sidebands for diversity or additional bandwidth, or up to four adjacent 3-kHz channels with independent transmissions on each channel). This permitted efficient sharing of the very limited spectrum available, and was appropriate for the services historically provided over HF channels: voice and low-speed data. However, recent years have seen increasing demand for higher-speed data transmission over HF links, and regulatory agencies are now examining the possibility of allocating single HF channels wider than 3 kHz. This concept is termed wideband HF (WBHF).

Global management of the radio spectrum resides with the International Telecommunications Union (ITU), an agency of the United Nations. Policies of the ITU are implemented by national administrations, such as the Federal Communications Commission (FCC) in the United States. Such agencies attempt to fairly balance the competing needs of all users of the spectrum. Typically, blocks of frequencies are reserved for each of the many services (e.g., fixed, mobile, aeronautical, and amateur) and special uses, such as radio astronomy. Channel widths (in hertz) vary over the electromagnetic spectrum, as well as within each band.

In the 3- to 30-MHz band (the nominal HF band), most channels for two-way communication (i.e., not broadcasting) are allocated only 3 kHz [1], although special uses (especially in the amateur bands) may be allocated narrower channels. Some military users have historically been allocated two (or even four) adjacent channels for independent sideband (ISB) operation. For example, the LINK-11 tactical data link sends the same information in two adjacent channels, achieving some extra robustness to fading when diversity combining at the receiver takes advantage of the imperfect correlation of the two channels.

6.2 The Need for Higher Data Rates

In this section, we present some applications wherein higher data rates could offer a qualitative improvement in mission performance. This discussion, drawn from [2], goes into enough detail to explore the application requirements and constraints for wideband HF waveforms.

6.2.1 Large Files to Fast Movers

Let's begin by examining the trade-offs of using higher data rate waveforms with an ARQ protocol. Consider an application in which aircraft pass within range of a transmitter for only a few minutes. The resulting hard limit on link duration will determine the maximum file size that can be sent, so higher data rates will allow larger files (e.g., images versus text) to be delivered as the aircraft flies through the radio range of the fixed transmitter.

While the raw speed of the data modem is of first-order importance, overhead–including link setup, link turnaround times, and the speed and robustness of ARQ acknowledgments–will result in diminishing returns as the modem speed increases.

As discussed in Chapter 5, we would like to use calling channels, separate from the traffic channels, for link setup. This tends to improve overall network efficiency because we can make heavy use of the traffic channels while keeping the calling channels open for new calls [3]. A narrowband (3 kHz) channel will suffice for link setup (see Chapter 7), which could take about ten seconds.

After link setup, the participating stations transfer data on a wideband traffic channel, using an automatic repeat request (ARQ) protocol. The interval between data bursts comprises the time to send the ACK plus two link turnaround times. Current modem and COMSEC technology requires about 1 second per link turnaround, which is an improvement over earlier equipment [4].

These relatively slow link turnarounds dominate the time to send an ACK, so efficiency will not suffer if we use a very robust, low-speed waveform for sending ACK bursts in a wideband channel. This could use multisymbol frames of orthogonal, Walsh-coded channel symbols, as is done in the 75-bps narrowband waveform in MIL-STD-188-110C (see Chapter 3) or STANAG 4415 [12].

6.2.2 Surveillance Video

A possible game-changing application for higher data rate waveforms is the delivery of real-time video streams over HF skywave channels. Video applications require data rates that are much higher than those currently available (at least 38 kbps) to carry even limited-quality imagery (e.g., 15 frames per second at a resolution of 160 × 120). While there are clearly applications where surface-wave transmission of video would be valuable, higher data rate waveforms must be able to operate in the fading and multipath distortion that is typical of long-haul skywave paths in order to provide these services over the widest possible range of conditions and users.

6.2.2.1 Long-Haul: Video from Aircraft

Today, unmanned aerial vehicle (UAV) commands and video are generally carried by satellite or line-of-sight radio channels. While an over-the-horizon command uplink to a UAV is feasible using a narrowband HF link, higher HF data rates offer the intriguing possibility of beyond line-of-sight communications both to and from a UAV via HF radio. Of course, powering an HF transmitter and achieving useful HF antenna efficiency on a UAV will be challenging.

The latter concerns are more easily addressed on manned fixed- and rotary-wing aircraft. The ability to provide real-time video from a helicopter beyond the horizon would be useful, not only to military forces, but also for disaster relief operations in remote locations. For example, the United States Coast Guard reported that a key impediment in managing the Deepwater Horizon oil spill in the Gulf of Mexico was shortcomings in real-time monitoring of the situation. The ability to send video from their helicopters would have helped the effort tremendously.

6.2.2.2 NVIS: Remote Video Observation Posts

Another challenging opportunity for using WBHF to deliver video arises in mountainous or dense urban terrain, where line-of-sight communications are blocked. Here, NVIS paths might be able to deliver video by going over the obstructions.

6.2.3 Common Operating Picture (Surface-Wave)

One-to-many communication is used for maintaining a common operating picture (COP) and supporting collaborative planning among vessels in a naval battle group. This application would certainly benefit from the higher data rates available from WBHF. Propagation via the relatively benign surface-wave channel over seawater provides extensive coverage with low pathloss, but nodes in this HF LAN must share the channel. With a token passing channel access protocol, the channel is in use continually for data, ACKs, and passing the token. Higher data rates would both decrease latency and increase the amount of data shared among the vessels.

6.3 Achieving Higher Data Rates

In Section 6.1, we described the current bandwidth allocations and regulations governing HF. In Section 6.2, we discussed applications that could benefit from higher data rates. Now let us investigate possible approaches for achieving higher user data rates.

There exist at least three possible approaches: the first is to increase data rates utilizing current 3-kHz allocations; the second is to increase data rates by using multiple 3-kHz contiguous or noncontiguous channels; and the third is by using wider contiguous bandwidth waveforms.

For this discussion, we will use a measurement related to SNR: the signal power-to-noise density ratio (SPNDR). SPNDR is the ratio of the total signal power to the noise power contained in a 1-Hz bandwidth; it can be very helpful when comparing waveforms with different noise bandwidths.

6.3.1 3-kHz Waveforms

Increasing the data rate of a waveform in a fixed 3-kHz bandwidth is quite straightforward. The only variables available are FEC and the modulation density. Recall from Chapter 3 that M-PSK constellations above 8-PSK become very power-inefficient. More power-efficient M-QAM constellations are better choices for increasing data rates. Every time an M-QAM constellation size is doubled, there is approximately a 3-dB cost in SNR, while the information contained in the constellation increases by a single bit. As a bandwidth efficient modulation, M-QAM follows the Shannon capacity curve in requiring exponentially more power to achieve linear increases in data throughput.

Table 6.1 shows the required SNR and SPNDR, the constellation, and the data rates achievable for an AWGN channel (for a BER = 10^{-4}) using the MDR waveforms discussed in Chapter 3 and extrapolating to 19,200 bps (i.e., doubling the current data rate of 9600 bps while using the same FEC, which is a rate 3/4 code).

As can be seen in Table 6.1, the cost of doubling the data rate from 9600 to 19200 is 18 dB. In addition to this high SNR price, both the required dynamic range of the entire system and the peak-to-average ratio (PAR) of the waveform increase as the constellation size increases, creating an even greater overall cost. Table 6.2 provides a sampling of the worst case PAR for the MDR waveforms. Furthermore, the higher order constellations are even more susceptible to multipath and fading, which makes it extremely unlikely that the 4096-QAM constellation would be able to handle anywhere near the same multipath and fading channel conditions as 64-QAM. Although other choices exist for achieving 19,200 bps

Table 6.1 Data Rate, Constellation Size, Required SNR, and SPNDR for 3-kHz Waveforms

Data Rate (bps)	*Constellation Size*	*SNR in 3 kHz (dB)*	*SPNDR (dB)*
3200	4-PSK	9	43.8
4800	8-PSK	12	46.8
6400	16-QAM	14	48.8
8000	32-QAM	17	51.8
9600	64-QAM	20	54.8
11,200	128-QAM	23	57.8
12,800	256-QAM	26	60.8
14,400	512-QAM	29	63.8
16,000	1024-QAM	32	66.8
17,600	2048-QAM	35	69.8
19,200	4096-QAM	38	72.8

Table 6.2 Worst-Case PAR for MDR Waveforms

Constellation	*PAR Before Radio Filter (dB)*	*PAR After Radio Filter (dB)*
4-PSK	2.4	4.9
16-QAM	4.0	5.7
64-QAM	5.1	6.6

without requiring 4096-QAM and such high SNRs, reasoning based on Shannon's capacity theorem suggest that the costs would similarly be very high. Undoubtedly, not allowing the bandwidth to increase beyond 3 kHz poses a significant challenge to the goal of realistically achieving higher data rates over HF.

In the last 10 years, there has been an enormous interest in multiple-input multiple-output (MIMO) systems as a way of increasing the data rates in a fixed bandwidth [5]. MIMO systems are also known as *space-time modulation* because of the added dimension of space that is created by the multiple antennas. The basic concept is to transmit the data over multiple transmit (TX) antennas and to receive the data using multiple receive (RX) antennas. Note that the receiver will require special MIMO processing to demodulate all the transmitted/received signal pairs simultaneously. If the number of TX antennas is N and the number of RX antennas is greater than or equal to N, the data rate of the system can be increased by the value N (assuming there is enough multipath in the system to support the N channels [6]). This technique has mostly been applied to systems operating in the 2 GHz or higher frequency range (where antenna spacing is small due to the shorter wavelengths).

In HF, MIMO techniques may be difficult to apply because of the longer wavelengths, which result in antennas that must be large in size. When spatial separation is used to decorrelate signals, this will require larger separation between all the TX antennas and all the RX antennas (although an argument can be made for using collocated antennas with different polarizations in this context).

In addition, the "enough multipath" constraint required in order to achieve the N-fold increase in data rate may not be available. On an AWGN channel, MIMO systems do not provide opportunities for increasing data rates (as described above) because a multipath rich environment is required [7]. In any case, it seems clear that, at best, MIMO would likely only be applicable to communications between a relatively small subset of HF sites with sufficient real-estate in order to be able to support multiple large HF antennas.

6.3.2 Multichannel Waveforms

An approach for increasing data rates over HF that fits nicely with current bandwidth allocations and existing radio equipment is the multichannel approach. The idea behind this approach is simply to use multiple 3-kHz channels in parallel. This approach offers users a linear increase in data rate as a function of the number of channels available. Modem implementers face only a linear increase in computational complexity for demodulating the multiple channels.

Let us illustrate this approach by comparing the performance of the 9600 bps waveform of MIL-STD-188-110B Appendix C [13] (labeled 110B/C) to the

9600-bps independent sideband (ISB) waveform (using 4800 bps per 3-kHz channel), which is defined in US MIL-STD-188-110B Appendix F [13] (labeled 110B/F). In order to compare the two approaches fairly, the total transmit power of both waveforms must be the same. In addition, since peak-power limited amplifiers (i.e., linear amplifiers) are used, the difference in the PAR of the waveforms must also be accounted for. For example, the SNR required by 110B/C to achieve a BER of 10^{-4} on the AWGN channel is 19 dB. For the 110B/F waveform, the SNR required per channel is 11 dB. This yields a gross advantage of 8 dB for using 110B/F instead of 110B/C. If two separate radios are used to transmit 110B/F and the total TX power is required to be the same, the advantage of 110B/F drops by 3 dB to 5 dB. However, accounting for the difference in PAR (64-QAM PAR is 6.6 dB, 4800 PAR is 4.9 dB), the advantage of 110B/F increases to 6.7 dB. If a single ISB radio is used instead of two radios, the PAR of 110B/F increases by 2 dB. This increase in PAR happens when both sidebands are combined before power amplification (note that the 2 dB value is an actual measured value); the advantage of 110B/F in an ISB radio is now 4.7 dB. If more than two sidebands are combined, PAR continues to increase (i.e., eight adjacent sidebands in a single radio would increase PAR by about 8 dB). Table 6.3 compares the SNR required for each waveform to achieve a BER of 10^{-4} on the three test channels of STANAG 4539. Even at 9600 bps, the benefits of the multichannel approach are significant.

An important assumption that was made for the 110B/F waveform in the previous results is that each channel was assumed to have the same average HF channel conditions and that the fading observed on each channel was independent of the other channel. This assumption may not be realistic for ISB channels because the fading observed on adjacent 3 kHz channels may not be completely uncorrelated. Furthermore, if the two channels are separated by 1 MHz (or more), the probability that both channels exhibit the same amount of multipath, fading, and SNR is very small. This assumption is the main drawback of US MIL-STD-188-110B Appendix F and other approaches like it [8] for obtaining higher data rates. These waveforms always use the same symbol constellation (i.e., 8-PSK, 16-QAM, etc) on all the available channels. They also spread the FEC and interleaving between all the channels. This requires that similar HF channel characteristics be observed on all channels for the waveform to function well. The system must discard channels that do not support a higher data rate available on other channel(s). In the worst case, this might result in only a single channel able to support the transmission with a good signal to noise ratio.

Other alternatives that could provide a more effective use of the multiple channels are: (1) develop a multichannel automatic repeat request (ARQ) protocol that

Table 6.3 Comparison of 110C/C and 110C/F 9600 bps Waveform

Channel	*110B/C (dB)*	*110C/F (dB per channel)*	*110C/F Gross Advantage (dB)*	*Advantage 2 Radios (dB)*	*Advantage ISB Radio (dB)*
AWGN	19	11	8	6.7	4.7
Poor	27	17	10	8.7	6.7
Rician	27	14	13	11.7	9.7

attempts to maximize the data rate of each individual channel and can thus achieve the highest possible multichannel data rate utilizing standard modems, waveforms and radio equipment; and (2) develop a multichannel STANAG 4538 (3G) ARQ protocol.

The practical drawbacks to using the multichannel approach in the field are that many 3 -Hz HF channel allocations are required in addition to many radios, antennas, modems, and so on. Although in recent years there has been much discussion of multichannel radios, the authors are not aware of any current developments of HF radios with more than two channels (except for strategic systems where the intent is to use each channel for a different application rather than to use all channels for a single application).

A final open question on the topic of multichannel waveforms is whether users are ready to pay the high price required to implement multichannel waveforms. Although the 110B/F waveform is being deployed in naval applications, due in large part to existing ISB frequency allocations and ISB radio equipment, it is unclear whether or not users that only have SSB equipment are willing to tie up a large portion of their radio assets on a single high data rate link.

6.3.3 Wider Contiguous Bandwidth Waveforms

Channel bandwidth plays a crucial role in the design and performance of high data rate waveforms. As was presented in Section 6.3.2, a two-channel ISB waveform (6 kHz) can have a significant performance advantage over a single-channel waveform (3 kHz) when providing a data rate of 9600 bps. As the data rate is decreased, this advantage becomes smaller. However, as the desired data rate increases beyond 9600 bps, this advantage can be even greater. Table 6.4 compares the required SPNDR for achieving a BER of 10^{-4} on a mid-latitude disturbed [9] channel for various data rates as a function of bandwidth for 110C/C waveforms and for a family of trellis-coded OFDM waveforms, as presented in [10]. Note that the PAR is not taken into account in this table, but would further increase the advantage of the wider bandwidth waveforms. What is astonishing about this comparison is that almost the same SPNDR is required to support 9600 bps in 3 kHz as is required

Table 6.4 Required SPNDR to Achieve a BER of 10^{-4} for ITU Mid-Latitude Disturbed Channel

Data Rate (bps)	*Bandwidth (kHz)*	*SPNDR (dB)*
9600	3	62
	12	55
19,200	3	> 78
	24	58
32,000	3	> 90
	40	60
64,000	3	> 100
	80	63

to support 64,000 bps in 80 kHz. Unmistakably, allowing the bandwidth of the waveform to increase yields much more power efficient waveforms.

6.3.4 Best Approach for WBHF

Of the three choices presented for increasing data rates over HF, the wider contiguous bandwidth approach offers the best performance. The PAR of the wider contiguous bandwidth waveforms would be the same as for 3-kHz waveforms; a multichannel approach implemented in a single radio would suffer from increased PAR due to the combining of the channels before the power amplifier. We therefore prefer the wider contiguous bandwidth approach.

The next decision is whether to use a single-carrier or OFDM waveform. As mentioned in Chapter 3, a possible drawback of single-carrier waveforms is the computational complexity of the adaptive equalizer. If we are to maintain the same delay spread capability, the number of taps required by the equalizer grows linearly with the increase in bandwidth, while equalizer complexity grows in proportion to the square of the number of taps. Although the computational complexity of the equalizer is a challenge for wider bandwidth waveforms, we have found that current digital signal processors (DSPs)—in combination with field-programmable gate arrays (FPGAs)—allow the implementation of single-carrier waveforms for bandwidths of at least 24 kHz. Considering the other drawbacks of OFDM waveforms (see Chapter 3), we prefer the single-carrier waveform as long as the equalizer remains feasible.

6.4 Standardization of Wideband HF Technology in the United States

One of the key new technologies added with the latest revision of the United States military standards for HF radios (MIL-STD-188-141C) and data modems (MIL-STD-188-110C) [13] is wideband HF (WBHF) waveforms for channels up to 24 kHz wide.

The modification to the radio standard was straightforward: the previous 3 kHz audio passband from 300 Hz to 3050 Hz was generalized. For a nominal bandwidth B, the audio frequency response is now required to have no more than 3-dB ripple from 300 Hz to $B + 50$ Hz.

Specification of WBHF waveforms necessarily requires much more detail. The WBHF waveforms are described in Sections 6.4.1, 6.4.2, and 6.4.3.

6.4.1 Design Goals

In establishing the design goals for the new family of wideband waveforms, it is important to understand the physical media in which they will be used (i.e., HF channel characteristics). For mid-latitude HF circuits, the amount of multipath (often called *delay spread*) can range up to 6 ms and the fade rate (often called *Doppler spread*) can be as high as 5 Hz. However, more typical values are 2-ms or less for delay spread, and 1 Hz or less for Doppler spread. The 2 ms delay spread and the 1-Hz Doppler spread values are the basic parameters of the standardized CCIR poor HF channel, or as it is now referred to in the latest ITU recommendation [11], the *mid-latitude disturbed channel condition*.

Besides the physical characteristics of HF, it is also important to understand the needs of the users and their applications. For some time now, many HF systems have used automatic repeat request (ARQ) protocols to provide error-free data delivery (e.g., FED-STD-1052 and STANAG 5066). Thus, waveforms need to be designed to work well with these applications, as well as traditional broadcast and point-to-point message handling systems.

Another application of HF has been to network together naval vessels and aircraft in close proximity to each other (extended line of sight, or about 200 nautical miles; see Section 6.2.3). This particular application is over surface-wave links that present much more benign channel conditions and much more consistent signal-to-noise ratios (SNRs) than are commonly encountered on skywave HF links. Higher data rates made possible by wider bandwidths lead to exciting possibilities in terms of the kinds of networking applications that can be supported. In particular, providing a variety of near real-time services over IP-based networks becomes practical with the data rates achievable with wideband HF.

Based on all of the above, the following three sets of designs goals steered the design of the MIL-STD-188-110C Appendix D waveforms.

6.4.1.1 General Design Goals

General design goals are:

- Bandwidths from 3 kHz through 24 kHz, in increments of 3 kHz.
- Increasing robustness as the data rate is decreased, not only with respect to reduced signal-to-noise ratio, but also with respect to delay and Doppler spreading.
- Variable length preamble allows for using a long preamble for skywave links, but a short preamble for surface-wave links. A long preamble is also valuable for longer transmissions; missing the preamble costs the user a lot in terms of wasted bandwidth. Short preambles are attractive for short transmissions that either can't afford the latency (voice) or function better with low latency, even at the cost of possible missed receptions (TDMA).
- Interleaver length not tied to preamble duration.
- Performance versus SNR for same modulation similar to US MIL-STD-188-110B 3 kHz waveforms (where bandwidth for computing SNR is the actual waveform bandwidth).
- Broadcast capability *without* autobaud on data (i.e., receiving modems must know all waveform parameters).
- Tail-biting FEC codes (similar to 110B) to reduce coding overhead.
- Streaming data rates compatible with most DTE equipment.

6.4.1.2 Skywave Design Goals

Skywave design goals are to have:

- Multipath capability of at least 6 ms.
- Long interleaver setting comparable to the long interleaver in 110C/C.
- Doppler spread capability of at least 8 Hz (for the lowest data rate waveforms in each bandwidth).
- Robust low rate modulation scheme providing performance similar to STANAG 4415 [12].

6.4.1.3 Surface-Wave Design Goals

Surface-wave design goals (i.e., highest data rate waveforms) are:

- Multipath capability of at least 3 ms in order to extend the range of surface-wave links to include reflections from the ionosphere (i.e., a Rician channel).
- Ultrashort interleaver setting (approximately 120 ms) and shortened preamble to support very low latency operation.
- End of transmission (EOT) marker in the on-air waveform to facilitate quick link turnarounds not requiring detection of data.
- High FEC code rates (i.e., rate 8/9, 9/10) to achieve the highest possible data rates.

6.4.2 WBHF Waveform Design

The design of the new wideband HF waveforms is similar to the MIL-STD-188-110B Appendix C waveforms [13] that were discussed in Chapter 3, but a few features were added that allow for a more flexible waveform.

We begin the discussion with the WBHF frame structure, shown in Figure 6.1.

As in the earlier waveforms, each transmission begins with a transmit level control (TLC) block. No information is carried by the TLC. It is only present to allow radio transmit gain control (TGC), transmitter automatic level control (ALC), and receiver automatic gain control (AGC) loops to settle before the actual preamble is sent/received. The length of the TLC section can be varied to suit the radios in use.

A variable-length preamble (see Section 6.4.4) follows the TLC section. This preamble is used for reliable synchronization and autobauding at the start of a transmission. The variable length feature allows a user to select the preamble length based on the expected channel conditions and the application. For example,

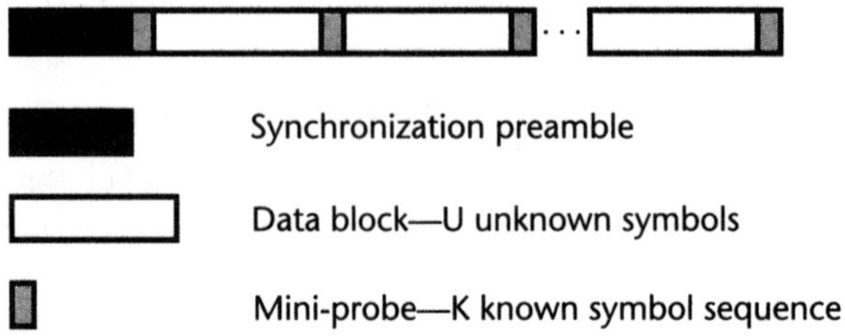

Figure 6.1 Frame structure (for waveforms 1–13).

a maritime surface-wave application may benefit from a very short preamble, whereas a very challenging multipath fading channel may benefit from a very long preamble. (The MIL-STD-188-110B Appendix C preamble has a fixed length.)

The preamble is followed by frames of alternating data (unknown) and probe (known) symbols[1].

6.4.3 WBHF Data Modulations

Eight bandwidths are available from 3 kHz through 24 kHz, in increments of 3 kHz. The subcarrier frequency and symbol rate for each bandwidth are listed in Table 6.5.

Each bandwidth offers up to 13 different data rates[2]. Modulations range from 2-ary phase shift keying (2-PSK) up to 256-ary quadrature amplitude modulation (256-QAM). The lowest data rate in each bandwidth (i.e., waveform ID (WID) 0) is based on the very robust STANAG 4415 Walsh modulation format [12].

A brief summary of the waveforms, including modulations and data rates (in bits per second), is presented in Table 6.6. The entries with a "-" are waveforms not used at certain bandwidths. The entries generally intended to provide surface-wave operation are WID 11 and 12.

6.4.3.1 PSK and QAM Waveforms

Waveforms 1–7 and 13 employ textbook PSK modulation, while waveforms 8–12 use specially modified QAM. As noted in Chapter 3, we can minimize the PAR of QAM waveforms by fitting the points within the unit circle, rather than employing the more typical square QAM constellations used in other media. The QAM constellations used for WBHF are shown in Figure 6.2.

Data symbols for Waveforms 1 through 7 and 13 (using BPSK, QPSK, or 8PSK modulation) are scrambled by modulo 8 addition with a pseudo-noise scrambling

Table 6.5 Modulation, Bandwidth and Data Rate

Bandwidth (kHz)	*Subcarrier (Hz)*	*Symbol Rate (Sym/sec)*
3	1800	2400
6	3300	4800
9	4800	7200
12	6300	9600
15	7800	12000
18	9300	14400
21	10800	16800
24	12300	19200

1. Waveform 0 is the exception; a Walsh modulation is used and no probe symbols are sent.
2. A fourteenth waveform is defined for 3-kHz bandwidth only. This is a new 2400-bps waveform that operates at lower SNR than the original 3-kHz, 2400-bps waveform.

Table 6.6 Modulation, Bandwidth and Data Rate

WID	*Modulation*	*3 kHz*	*6 kHz*	*9 kHz*	*12 kHz*	*15 kHz*	*18 kHz*	*21 kHz*	*24 kHz*
0	Walsh	75	150	300	300	300	600	300	600
1	2-PSK	150	300	600	600	600	1200	600	1200
2	2-PSK	300	600	1200	1200	1200	2400	1200	2400
3	2-PSK	600	1200	2400	2400	2400	4800	2400	4800
4	2-PSK	1200	2400	-	4800	4800	-	4800	9600
5	2-PSK	1600	3200	4800	6400	8000	9600	9600	12,800
6	4-PSK	3200	6400	9600	12,800	16,000	19,200	19,200	25,600
7	8-PSK	4800	9600	14,400	19,200	24,000	28,800	28,800	38,400
8	16-QAM	6400	12,800	19,200	25,600	32,000	38,400	38,400	51,200
9	32-QAM	8000	16,000	24,000	32,000	40,000	48,000	48,000	64,000
10	64-QAM	9600	19,200	28,800	38,400	48,000	57,600	57,600	76,800
11	64-QAM	12,000	24,000	36,000	48,000	57,600	72,000	76,800	96,000
12	256-QAM	16,000	32,000	45,000	64,000	76,800	90,000	115,200	120,000
13	4-PSK	2400	—	—	—	—	—	—	—

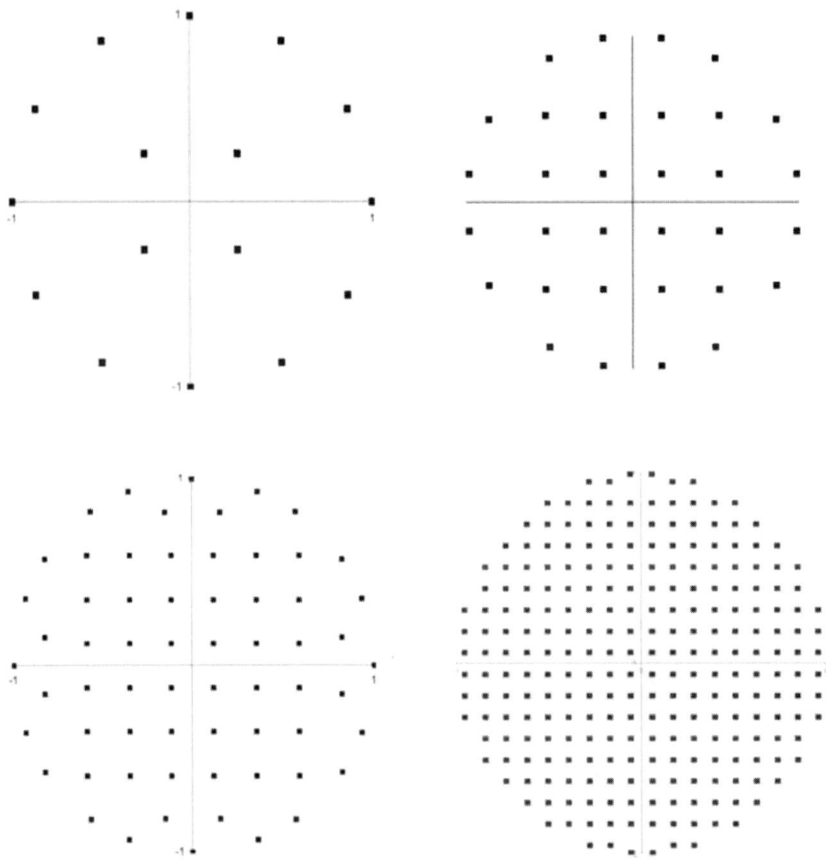

Figure 6.2 QAM constellations used in WBHF waveforms.

sequence (described below). This results in all of these waveforms appearing to be 8 PSK on the air.

The data symbols for Waveforms 8 through 12 (16QAM, 32QAM, 64QAM and 256QAM) are scrambled using an exclusive or (XOR) operation. The data bits for each symbol (4, 5, 6, or 8 bits) are XOR'd with an equal number of bits from the scrambling sequence.

For Waveforms 1 through 13, the scrambling sequence generator polynomial is $x^9 + x^4 + 1$, as shown in Figure 6.3. In this illustration, three output bits are shown; this is the case for all of the PSK waveforms. For 2^N QAM waveforms, the rightmost N bits are used. The generator is initialized to 1 at the start of each data frame.

After each data symbol is scrambled, the generator is shifted the required number of times to produce all new bits for use in scrambling the next symbol (i.e., three iterations for any of the PSK waveforms, four iterations for 16-QAM, and so on). Because the generator is iterated *after* the bits are used, the first data symbol of every data frame is always scrambled using the appropriate number of bits from the initialization value of 00000001.

The length of the scrambling sequence is 511 bits. For a 256-symbol data block with 6 bits per symbol, for example, the scrambling sequence will be repeated slightly more than three times. However, in terms of symbols there will be no repetition, since 511 is relatively prime to 3, 4, 5, 6, and 8.

6.4.3.2 Walsh Data Modulation

Waveform 0 uses a different modulation technique, *Walsh orthogonal modulation*. This is similar to the Walsh modulation used in the very robust 75-bps waveform in MIL-STD-188-110 and STANAG 4415 (see Chapter 3, Section 3.2.2.2). For each

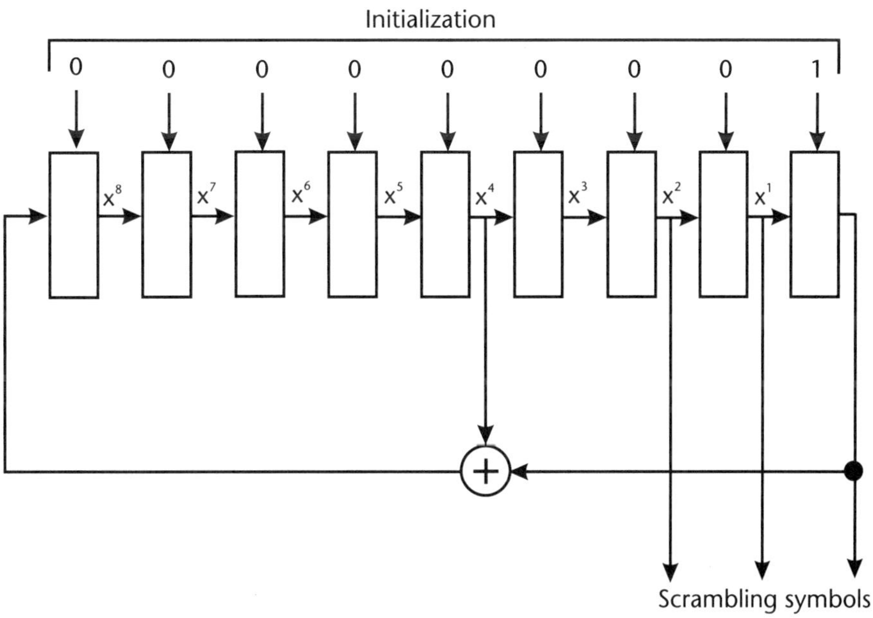

Figure 6.3 PSK/QAM scrambling sequence generator.

pair of coded and interleaved data bits, the method produces a 32-symbol repeated Walsh sequence. The Walsh orthogonal modulation is produced by taking each pair of bits, or dibit, and selecting a corresponding Walsh sequence from the second column of Table 6.7.

The selected four-element Walsh sequence is repeated 8 times to yield a 32-element Walsh sequence. For example, if the dibit is 01, the sequence 0404 is repeated to generate

$$0, 4, 0, 4, 0, 4, 0, 4, 0, 4, 0, 4, 0, 4, 0, 4, 0, 4, 0, 4, 0, 4, 0, 4, 0, 4, 0, 4, 0, 4, 0, 4$$

However, the last dibit in any interleaver block is distinguished by using the set of *Alternate Walsh Sequences* (shown in the last column in Table 6.7), which are repeated four times to produce a 32-element sequence.

The 32-element channel symbol in each case is produced by an element-by-element modulo 8 addition of the repeated Walsh sequence with a special 8-PSK scrambling sequence.

6.4.4 Synchronization Preamble

The synchronization preamble is used for rapid initial synchronization and provides time and frequency alignment. The preamble consists of two main sections, a transmitter level control (TLC) settling time section, followed by a synchronization section that contains a repeated preamble superframe (see Figure 6.4).

Since we want the sync preamble to be very robust, we use the same 4-ary orthogonal Walsh modulation used in Waveform 0 (from the first two columns of Table 6.7). The length of each channel symbol, in chips or symbols, is dependent on the bandwidth of the modem waveform selected: 32 symbols for each 3 kHz of waveform bandwidth[3]. The expanded Walsh sequences are scrambled using a distinct scrambling sequence for each subsection of the preamble: fixed, count, and waveform ID.

6.4.4.1 TLC Section

The first section of the preamble, denoted TLC, is provided exclusively for radio and modem TGC and AGC, and consists of *N* blocks of 8-PSK. The length of each

Table 6.7 Walsh Functions

Di-bit	*Walsh Sequence*	*Alternate Walsh Sequence*
00	0000	00004444
01	0404	04044040
10	0044	00444400
11	0440	04404004

3. Thus, the duration of each Walsh channel symbol is 13.3 ms, independent of bandwidth.

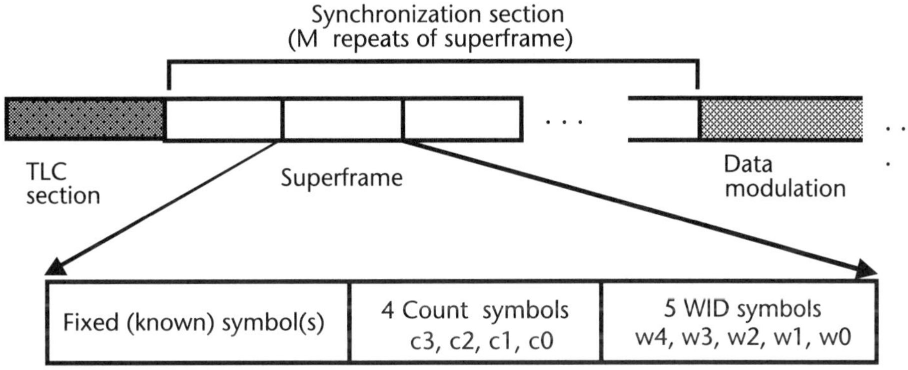

Figure 6.4 Synchronization preamble structure.

block in PSK symbols depends on the bandwidth used in the transmission to follow. The value of N is configurable to range from 0 to 255 (for $N = 0$, the TLC section is not transmitted).

6.4.4.2 Synchronization Section

The synchronization section of the preamble contains a repeated preamble superframe (Figure 6.4). The preamble superframe consists of three distinct subsections, one with a fixed (known) modulation, one to convey a downcount, and one to convey waveform identification. The superframe is repeated M times. The synchronization section is immediately followed by the modulated data (Figure 6.4).

Fixed Subsection

The *fixed subsection* of each super-frame consists of either 1 or 9 orthogonal Walsh modulated channel symbols, with 9 symbols being the normal case and the single fixed symbol used only in conjunction with a shortened preamble where the superframe is not repeated. The length of each channel symbol depends on the bandwidth. For the case of the single Walsh symbol the dibit is 3 (binary 11), and the superframe is transmitted only once ($M = 1$). For the case of 9 Walsh symbols the dibit sequence is 0, 0, 2, 1, 2, 1, 0, 2, 3. The fixed subsection is intended exclusively for synchronization and Doppler offset removal purposes.

Preamble Downcount Subsection

The next subsection consists of four orthogonal Walsh modulated dibits, labeled as c_3, c_2, c_1, and c_0, each conveying two bits of information. This subsection represents a 5-bit downcount plus 3 parity bits. This count is initialized to a value of (M-1) and is decremented with each of the M preamble repetitions until it reaches zero in the final super-frame before data begins.

The 5-bit superframe down count is initialized to M-1 where M is the number of repeats of the superframe. The bits in the binary representation of the downcount are labeled $b_4b_3b_2b_1b_0$, where b_4 is the MSB and b_0 is the LSB. Bits b_7, b_6, and b_5 contain a parity check computed over $b_4b_3b_2b_1b_0$ as follows, where the ^ symbol indicates exclusive-or:

$$b_7 = b_1 \wedge b_2 \wedge b_3$$

$$b_6 = b_2 \wedge b_3 \wedge b_4$$

$$b_5 = b_0 \wedge b_1 \wedge b_2$$

C_3 contains the two MSBs b_7 and b_6, c_2 contains the next two bits b_5 and b_4, and so on.

Waveform ID Subsection

The final subsection of the preamble superframe consists of five orthogonal Walsh modulated channel symbols, each conveying two bits of information. These dibits are labeled as w_4, w_3, w_2, w_1, and w_0. These 10 bits represent a Waveform ID consisting of waveform number, interleaver option, convolutional code length, and parity check. The 10 bits are labeled d_9 down to d_0. W_4 contains d_9 and d_8, whereas d_9 is the MSB, w_3 contains d_7 and d_6, and so on down to w_0, which contains d_1 and d_0.

- The 4-bit waveform number is encoded in the w_4 and w_3 dibits as a binary number. (Values greater than 13 are reserved.)
- The Interleaver selection is encoded in w_2 as a binary number, 0 for the ultra-short interleaver through 3 for the long interleaver. (Interleaving is described in Section 6.4.4.)
- The constraint length of the convolutional code is encoded in the msb of w_1: 0 for constraint length 7, 1 for constraint length 9. The lsb of w_1 is always 0.

The 3 LSBs, d_2, d_1, and d_0, contain a 3-bit checksum calculated over $d_9d_8d_7d_6d_5d_4d_3$ as follows, where the ^ symbol indicates exclusive-or:

$$d_2 = d_9 \wedge d_8 \wedge d_7$$

$$d_1 = d_7 \wedge d_6 \wedge d_5$$

$$d_0 = d_5 \wedge d_4 \wedge d_3$$

6.4.5 Data Blocks and Miniprobes

Each data frame consists of a data block of *U* (unknown) data symbols, followed by a miniprobe consisting of *K* known symbols. The known symbols are used to track the time-varying multipath channel, and the unknown symbols carry the user data (after encoding and interleaving). *U* and *K* vary with the bandwidth and data rate in use; example values for 3 kHz and 12 kHz bandwidths are listed in Table 6.8.

Miniprobes are inserted following every data block, and at the end of the preamble for non-Walsh based modulations (i.e., all except Waveform ID 0). To

Table 6.8 Data Block and Miniprobe Lengths for 3 kHz and 12 kHz

		3 kHz			*12 kHz*		
WID	*Modulation*	*Data rate*	*U*	*K*	*Data Rate*	*U*	*K*
1	BPSK	150	48	48	600	192	192
2	BPSK	300	48	48	1200	192	192
3	BPSK	600	96	32	2400	384	128
4	BPSK	1200	96	32	4800	384	128
5	BPSK	1600	256	32	6400	1024	128
6	QPSK	3200	256	32	12,800	1024	128
7	8PSK	4800	256	32	19,200	1024	128
8	16QAM	6400	256	32	25,600	1024	128
9	32QAM	8000	256	32	32,000	1024	128
10	64QAM	9600	256	32	38,400	1024	128
11	64QAM	12,000	360	24	48,000	1080	72
12	256QAM	16,000	360	24	64,000	1080	72
13	QPSK	2400	256	32	—	—	—

support the wide range of bit-rate and bandwidth options of this standard, 14 different miniprobe sequences are utilized. Each of the miniprobes consists of a base sequence cyclically extended to the required length.

The miniprobes are also utilized to identify the long interleaver block boundary. This is accomplished by transmitting a cyclically rotated version of the miniprobe following the second to last data block of the long interleaver frame. The position of this cyclically shifted miniprobe remains constant regardless of which interleaver has actually been selected. As all interleavers line up on the long interleaver block boundary, this feature can be used to synchronize to a broadcast transmission and provide a late entry feature when the Waveform ID fields are known in advance by the receiver. The cyclically rotated version of the miniprobe is obtained by first cyclically extending the base sequence, and then shifting by a predetermined number of symbols.

Table 6.9 defines the miniprobe lengths and the base sequence used to generate the full miniprobe and also the cyclic shift utilized to signal the interleaver block boundary.

6.4.6 Interleaving

Four interleaver sizes are available for WIDs 1 to 13 (Table 6.10). Only three of these are available for WID 0 (short, medium, and long). The smallest interleaver size will span approximately 120 ms, and each larger interleaver size will be four times the length of the previous interleaver. The interleaver is designed to separate neighboring bits in the coded data block as far as possible over the span of the interleaver, with the largest separations resulting for the bits that were originally closest to each other.

Table 6.9 Miniprobe Lengths and Base Sequences

Miniprobe Length	*Base Sequence*	*Cyclic Shift for Interleaver Boundary*
24	13	6
32	16	8
36	19	9
48	25	12
64	36	18
68	36	18
72	36	18
96	49	24
128	64	32
144	81	40
160	81	40
192	100	50
224	121	60
240	121	60
272	144	72

Table 6.10 Interleaver Options

Interleaver	*Length (s)*
Ultra-short (US)	~ 0.12
Short (S)	~ 0.48
Medium (M)	~ 1.92
Long (L)	~ 7.68

The block interleaver consists of a single dimension array starting at index 0 up to index N-1 (where N is the interleaver size in bits). Bit n is loaded into the interleaver by using the following equation

$$\text{Load Location} = (n * \text{Interleaver\_Increment\_Value}) \text{ Modulo } (N) \tag{6.1}$$

The Interleaver_Increment_Value in (6.1) is selected such that bit soft decisions at the input to the FEC decoder are fairly balanced (i.e., adjacent bits after deinterleaving are not the same bit location in M-PSK or M-QAM constellations). An example set of increment values (for the 12-kHz waveforms) is listed in Table 6.11.

The increment values were chosen to ensure that the combined cycles of puncturing and that the assignment of bit positions in each symbol for the specific constellation being used is the same as if there had been no interleaving. For waveforms 7 to 12, this is important, because each symbol of a constellation contains *strong* and *weak* bit positions. A *strong bit position* is one that has a large average distance between all the constellation points where the bit is a 0 and the closest point where it is a 1. Typically, the MSB is a strong bit and the LSB a *weak bit*

Table 6.11 Interleaver Increment Values for 12-kHz Waveforms

	Interleaver Size			
WID	*Ultrashort*	*Short*	*Medium*	*Long*
0	—	41	145	577
1	73	289	1153	4633
2	73	289	1153	4633
3	97	385	1537	6193
4	97	385	1537	6193
5	129	513	2049	8321
6	257	1025	4097	16641
7	385	1537	6145	24961
8	513	2049	8193	33281
9	641	2561	10241	41603
10	769	3073	13057	49921
11	811	3241	13771	52651
12	1081	4321	18361	70201

position. An interleaving strategy that does not evenly distribute these bits in the way they occur without interleaving could degrade performance.

An additional constraint on the increment values is that, when possible, adjacent bits after deinterleaving must be separated by several alternating blocks of known/unknown frames over the air. The larger the interleaver size, the larger this separation can be made. This constraint helps improve performance on slowly fading channels.

6.4.7 FEC

Iterative codes were not considered for the new wideband HF waveforms due to the continued requirement that the standard be free of any intellectual property (turbo codes are patented technology). Thus, the coding choice for the new wideband waveforms was to use the standard rate 1/2 constraint length 7 convolutional code that has been used for over two decades in 110A and 110B, with the addition of a constraint length 9 code to provide additional coding protection at a cost of increased computational complexity. For additional versatility, repetition coding and puncturing were used to create a wide range of coding options in order to achieve the data rates shown in Table 6.6.

Very high code rates (i.e., 8/9, 9/10) are used to attain the highest data rates (for surface-wave links). However, very high puncturing of convolutional codes can result in very weak codes. The optional constraint length 9 code was added because it is a much stronger code when highly punctured. Users of the WBHF standard have the option of selecting a $k = 7$ or a $k = 9$ code.

Table 6.12 provides the code rates that are used for each modulation and bandwidth. Table 6.13 provides the puncturing and repetition patterns. Note that the puncturing patterns can be also found in [14]. The entries with a "-" once again indicate combinations that are not used.

Table 6.12 Modulation, Bandwidth, and Code Rate

WID	*3 kHz*	*6 kHz*	*9 kHz*	*12 kHz*	*15 kHz*	*18 kHz*	*21 kHz*	*24 kHz*
0 - Walsh	1/2	1/2	2/3	1/2	2/5	2/3	2/7	1/2
1 - 2-PSK	1/8	1/8	1/8	1/8	1/12	1/8	1/16	1/8
2 - 2-PSK	1/4	1/4	1/4	1/4	1/6	1/4	1/8	1/4
3 - 2-PSK	1/3	1/3	1/2	1/3	1/3	1/2	1/4	1/3
4 - 2-PSK	2/3	2/3	—	2/3	2/3	—	1/2	2/3
5 - 2-PSK	3/4	3/4	3/4	3/4	3/4	3/4	2/3	3/4
6 - 4-PSK	3/4	3/4	3/4	3/4	3/4	3/4	2/3	3/4
7 - 8-PSK	3/4	3/4	3/4	3/4	3/4	3/4	2/3	3/4
8 - 16-QAM	3/4	3/4	3/4	3/4	3/4	3/4	2/3	3/4
9 - 32-QAM	3/4	3/4	3/4	3/4	3/4	3/4	2/3	3/4
10 - 64-QAM	3/4	3/4	3/4	3/4	3/4	3/4	2/3	3/4
11 - 64-QAM	8/9	8/9	8/9	8/9	8/9	8/9	4/5	8/9
12 - 256-QAM	8/9	8/9	8/9	8/9	8/9	5/6	9/10	5/6
13 – 4-PSK	9/16							

Table 6.13 Puncture and Repetition Patterns

Code Rate	*K = 7* *Puncture Pattern*	*K = 9* *Puncture Pattern*	*Number of Repeats*
9 / 10	111101110 100010001	111000101 100111010	n/a
8 / 9	11110100 10001011	11100000 10011111	n/a
5 / 6	11010 10101	10110 11001	n/a
4 / 5	1111 1000	1101 1010	n/a
3 / 4	110 101	111 100	n/a
2 / 3	11 10	11 10	n/a
9/16	111101111 111111011	111101111 111111011	n/a
1 / 2	n/a	n/a	n/a
2 / 5	1110 1010	1110 1010	1/2 repeated 2x
1 / 3	n/a	n/a	2/3 epeated 2x
2 / 7	1111 0111	111 0111	1/2 repeated 2x
1 / 4	n/a	n/a	1/2 repeated 2x
1 / 6	n/a	n/a	1/2 repeated 3x
1 / 8	n/a	n/a	1/2 repeated 4x
1 / 12	n/a	n/a	1/2 repeated 6x
1 / 16	n/a	n/a	1/2 repeated 8x

6.4.8 Standardized Feature Packages

To help structure the marketplace for WBHF modems, the standard specifies three blocks of features:

- *WBHF Block 1* includes all of the waveforms but only in a 3-kHz channel, and only the constraint length 7 FEC. The principal advantage of a WBHF Block 1 modem over previous-generation 3-kHz HF modems is the improved 2400-bps waveform. You also get a robust 1600-bps waveform and new surface-wave waveforms at 12 kbps and 16 kbps.
- *WBHF Block 2* includes all of the waveforms at bandwidths of 3, 6, 9, and 12 kHz. It offers considerably higher data rates than previous HF data modems to users who can obtain allocations of up to 12 kHz. As in Block 1, the FEC is constraint length 7 only.
- *WBHF Block 3* includes all of the WBHF capabilities: all waveforms, both FEC modes for all waveforms, and bandwidths up to 24 kHz.

6.4.9 WBHF Performance Requirements

The similarity of the WBHF modulations to the previous-generation HF data modems leads to similar performance capabilities in SNR terms. (Remember that achieving the same SNR in a channel with twice the bandwidth requires twice as much power.) As a result, the performance requirements for the WBHF waveforms can be presented in a single table, independent of bandwidth (see Table 6.14). The few exceptional cases are noted (for example, where some waveforms are not available in some bandwidths).

The AWGN channel represents a surface-wave channel with a single, nonfading path. To obtain reliable measurements, each condition must be measured for at least 60 minutes.

The poor channel represents a skywave channel, with two independent but equal average power Rayleigh fading paths, with a fixed 2-ms delay between paths, and with a fading (two sigma) bandwidth of 1 Hz (i.e., the ITU-R mid-latitude disturbed channel). To obtain reliable measurements, each condition must be measured for at least five hours.

6.5 WBHF Application Performance

We now return to the WBHF applications introduced in Section 6.2, and estimate the performance of the new WBHF waveforms using analysis and simulation. Johnson [2] presented estimates of performance of a hypothetical 12-kHz WBHF modem before the WBHF standard was written. Here, we present a similar analysis, but using the actual required performance of the 12- and 24-kHz modems. SNR requirements for selected data rates in 12- and 24-kHz channels are listed in Table 6.15.

Table 6.14 WBHF Performance Requirements

	Average SNR (dB) for BER ≤ 1.0E-5		Exceptions
WID	*AWGN Channel*	*Poor Channel*	*(for Specific Bandwidths)*
0	−6	−1	9-kHz Poor channel only: allow +1 dB
1	−3	3	
2	0	5	
3	3	7	9 kHz: allow extra 1 dB (both channels)
4	5	10	This waveform not available in 9 or 18 kHz
5	6	11	
6	9	14	
7	13	19	
8	16	23	
9	19	27	
10	21	31	24-kHz Poor channel only: test for BER ≤ 1.0E-4 and allow 33 dB
11	24	—	
12	30	—	
13	6	11	This waveform available in 3 kHz only

Table 6.15 SNR Requirements for 12- and 24-kHz Waveforms

	SNR for 10-5 BER		Data Rate	
WID	*Surface-Wave*	*Skywave*	*12 kHz*	*24 kHz*
12	30	—	60,000	120,000
10	21	31	38,400	76,800
9	19	27	32,000	64,000
7	13	19	19,200	38,400
6	9	14	12,800	25,600
4	5	10	4,800	9,600
2	0	5	1,200	2,400
0	−6	−1	300	600

6.5.1 Estimated Application Performance: File Transfer

Our first application evaluates the benefit of WBHF in transferring files within a bounded time. Here, a unit on the ground uploads a file to an aircraft flying rapidly through the area. An extended-line-of-sight (ELOS) link is established as soon as the aircraft is within range. During the file upload opportunity, the SNR may vary by 100 dB (limited by capabilities of the radio) from the link establishment (minimum SNR) through the aircraft passing overhead (maximum SNR); the SNR will then fall again as the aircraft flies away from the transmitter. Therefore, we employ the STANAG 5066 ARQ protocol to adapt the data rate during the flyover.

For a fair evaluation of using the wideband waveforms, we compare the amount of data that can be sent during the transit of the aircraft *using a fixed RF power output* (10 to 1000 W) with either 3-, 12-, or 24-kHz waveforms. Because a fixed power output will produce a higher SNR in a narrower bandwidth, the narrowband system will have an SNR advantage at all ranges and may be able to pass data at ranges beyond that of the WBHF systems.

Two sccnarios arc modeled:

- First we consider a *fixed-base scenario*. The antenna is 30 feet above the ground, sending files to an aircraft flying at 500 knots at 30,000 feet above ground level (AGL). In this scenario, the narrowband transmitter is able to link with the aircraft at distances ranging from 100 to 250 miles, depending on the transmitter power. However, the wideband system must wait until the aircraft gets closer (e.g., 75 miles when the narrowband system could start sending data at 100 miles out). In Figure 6.5, we see that the wideband systems are able to send substantially more data in the finite flyover time than the narrowband system. The ratio ranges from 2.6 to 2.9 times more data for the 12-kHz system, and 3.8 to 4.5 times the file size for the 24-kHz system.
- Now, consider a hasty setup with a tactical antenna erected only 10 feet above the ground, and the aircraft flying at 600 knots at 5,000 feet AGL. In this *low and fast scenario*, the time available for the file upload is reduced significantly. The lower aircraft altitude reduces the range at which a link can be established, and the higher speed reduces the time that the aircraft is within that range. The narrowband transmitter is able to link with the aircraft at distances ranging from 30 to 75 miles, while the wideband systems link at 17 to 50 miles. Again, despite the shorter range and the reduced SNR of the

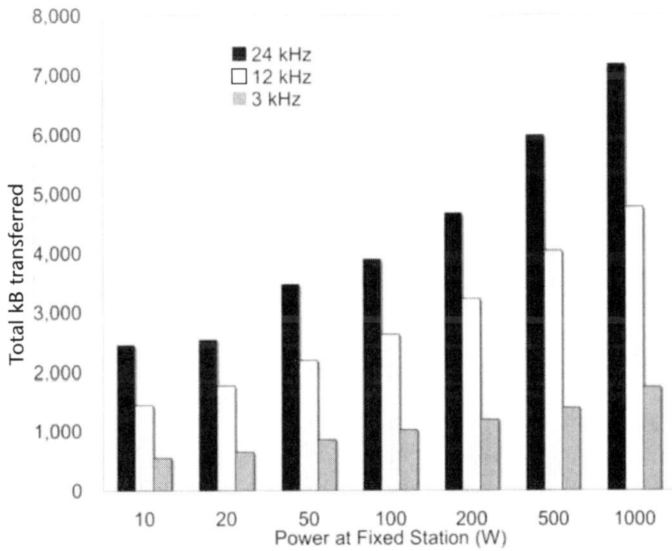

Figure 6.5 Fixed-base scenario file transfer.

wideband systems, their higher peak throughput is sufficient to provide 2.4 to 4.3 times as much data to the aircraft as it flies through the theater (see Figure 6.6).

6.5.2 Estimated Application Performance: Video Over HF Skywave

In our video application, we again consider two scenarios: a UAV delivering video via a long-haul skywave link, and a ground-based observation post using an NVIS path to transmit real-time video to observers one valley away. In both cases, we send H.264-compressed video streams (15 frames per second at a resolution of 160 × 120), using waveforms robust enough for skywave channels (i.e., not WID 11 or 12). We noted during over-the-air testing of video over WBHF (see Chapter 6, Section 6.6.2.1) that data rates below 19,200 bps are not very useful for carrying real-time video.

We assume that the H.264 compression application slices the video stream to match the packet size used by the MAC layer (e.g., 300 bytes) so that packet losses do not result in severe corruption of the video stream. Using a 1% packet loss rate as a threshold for marginal video quality, we set our bit error rate (BER) threshold at 3×10^{-6}. (This is quite conservative because it ignores the characteristic burstiness of errors in skywave channels.) The resulting SNR thresholds for various 12 and 24-kHz waveforms suitable for skywave video are listed in Table 6.16, along with the corresponding SPNDRs for 12- and 24-kHz channels.

6.5.2.1 UAV Scenario

While sending a video from UAV via WBHF is an exciting prospect, we must first overcome a number of challenges that are not directly related to the channel

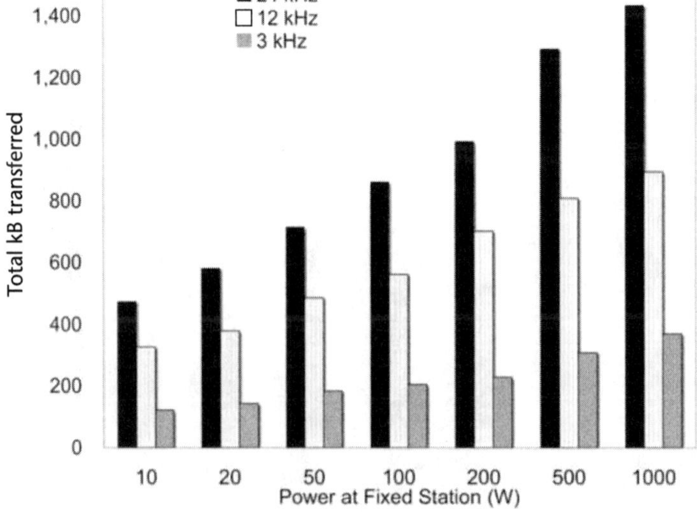

Figure 6.6 Low and fast scenario file transfer.

Table 6.16 SNR Thresholds for 12 and 24 kHz Waveforms at BER ≤ 3E-6

	Data Rate			*Estimated SPNDR for 3E-6 BER*	
WID	*12 kHz*	*24 kHz*	*Estimated SNR for 3E-6 BER*	*12 kHz*	*24 kHz*
10	38,400	76,800	34	75	78
9	32,000	64,000	29	70	73
7	19,200	38,400	22	63	66
6		25,600	17		61

bandwidth. In particular, the radiation efficiency of small aircraft is poor below about 9 MHz, with about 10 dB loss at 6 MHz and increasing loss below that [15]. Also, the payload power available on smaller UAVs may be insufficient for long-haul HF links. However, for medium-to-large UAVs, it may be practical to contemplate downlinking video via HF at ranges of a thousand miles or more.

To evaluate the feasibility, we analyzed a 1-kW HF transmitter on a UAV with a 0 dBi antenna. The receiver is 1515 km distant, using a horizontal log periodic antenna. In Figure 6.7, we show the SNR density for this scenario (from VOACAP) in the month of June with a smoothed sunspot number (SSN) of 55.

We see that, for this scenario, there are frequencies throughout the day that provide at least 65 dB/Hz. Referring to Table 6.16, a 12-kHz WBHF system would be able to send video at 19.2 kbps all day, with higher data rates for much of the day. If a 24-kHz WBHF system was used, we would have at least 25.6-kbps video quality, with even higher-quality video at 38.4 and even 64 kbps for much of the

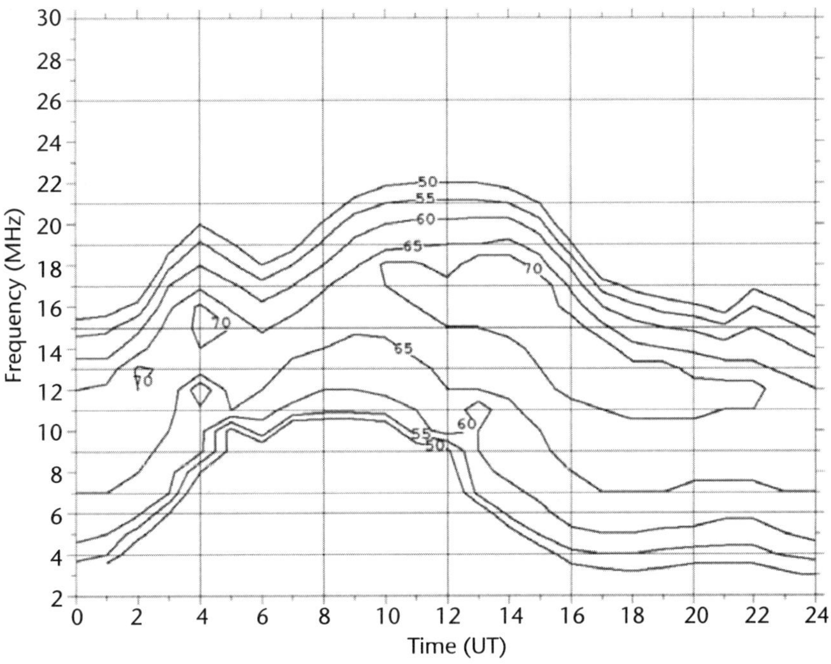

Figure 6.7 SNR Density for a 1515 km path, June, SSN 55, 1-kW transmitter.

day. Thus, downlinking UAV video over long-haul HF skywave channels appears feasible.

6.5.2.2 NVIS Scenario

For another application of delivering real-time video over WBHF, consider an unmanned observation post sited one valley away from the users. A ridge between the observation post and the users prevents line-of-sight communications, and requires that an NVIS HF path be used. We place a 400 W tactical WBHF radio and a horizontal dipole antenna at the observation post, and employ a horizontal yagi antenna at the receiver site. The VOACAP prediction for SPNDR on this path (again in June with SSN of 55) is shown in Figure 6.8.

Once again, we find that at least 65-dB SPNDR is available in every hour on the NVIS path. This suggests that we should be able to deliver acceptable quality real-time video over NVIS paths, as well as long-haul paths when the circuit is well-designed.

6.5.3 Estimated Application Performance: Common Operating Picture

We draw our example surface-wave scenario from the naval battle group LAN discussion in Chapter 3, Section 3.4: a group of six naval vessels shares a (wideband) channel using token passing. However, instead of the lower data rate applications considered there, we now investigate the ability of the surface-wave LAN to sustain higher bandwidth messaging to maintain a consistent common operating picture (COP). The higher data rates afforded by the wideband HF channel allow the use of general-purpose IP networking. However, the overheads associated with some

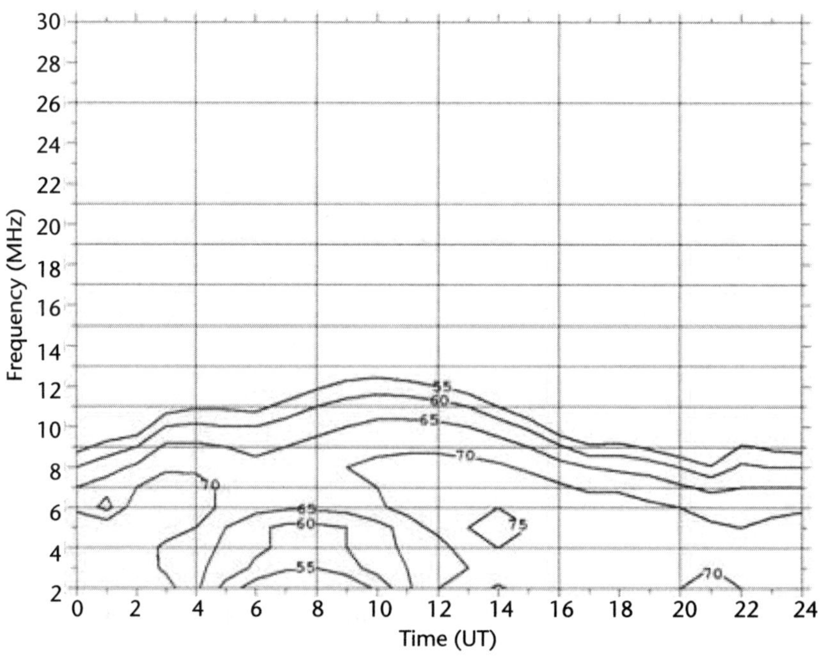

Figure 6.8 SNR Density for NVIS path, June, SSN 55, 400 W transmitter.

IP applications may be impractical over HF bearers, as noted in Chapter 5, Section 5.8.

In this application, battle group members exchange information with other nodes as the token circulates. The maximum token tenure is arbitrarily set to 9.6 s (this includes time to send ACKs and the token), optionally preceded by IP data (such as COP), and sent at 64 or 120 kbps (for the 12- and 24- kHz channels, respectively). The link turnaround time is 1 s.

We assume that one ship (designated Node A) receives a COP downlink via SATCOM and pushes a filtered subset of this data stream to the rest of the battle group via the WBHF LAN. Node A therefore sends data packets each time it receives the token. The other ships do not always have data to send; the fraction of transmit opportunities that they use will be varied in this experiment.

Two cases are considered for channel access by the other ships: (1) *Polling*: Node A receives half of the token tenures, alternating with the other ships, and (2) *Peer-to-peer*: Node A receives one token tenure per token rotation.

Our performance metric is total throughput in the surface-wave ELOS LAN, as a function of the fraction of transmit opportunities used by the ships other than Node A.

Three systems are compared: (1) a current 2-ISB (6-kHz) system with 19.2 kbps modems; (2) a 12-kHz WBHF system with 64-kbps modems; and (3) a 24-kHz WBHF system with 120-kbps modems.

In the polling case (Figure 6.9), we see that the 12 kHz WBHF system increases throughput more than threefold when compared to the 2-ISB case, despite using only twice the bandwidth. This is the direct result of the more aggressive surface-wave waveform (WID 12) introduced for the WBHF generation of modems, with its higher-rate FEC and 256 QAM constellation. The 24-kHz system achieves more than six-fold improvement over the current 2-ISB system.

We see very similar results in the peer-to-peer scenario (Figure 6.10), but with generally higher throughputs because we lose less time to link turnarounds. In the

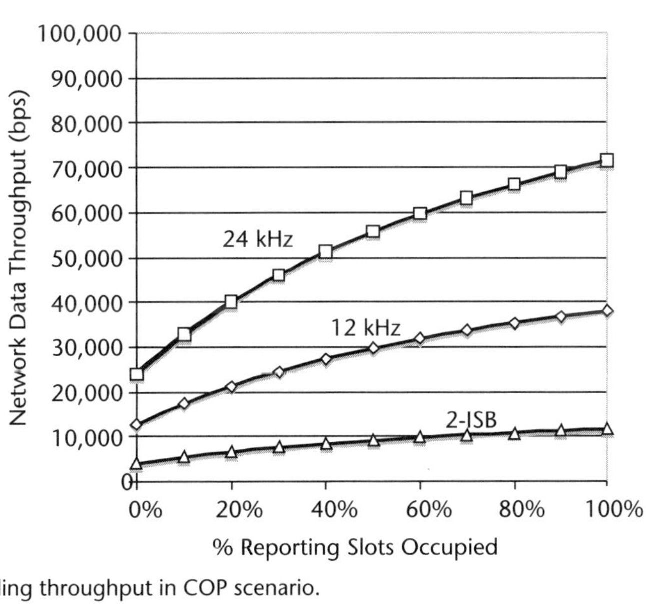

Figure 6.9 Polling throughput in COP scenario.

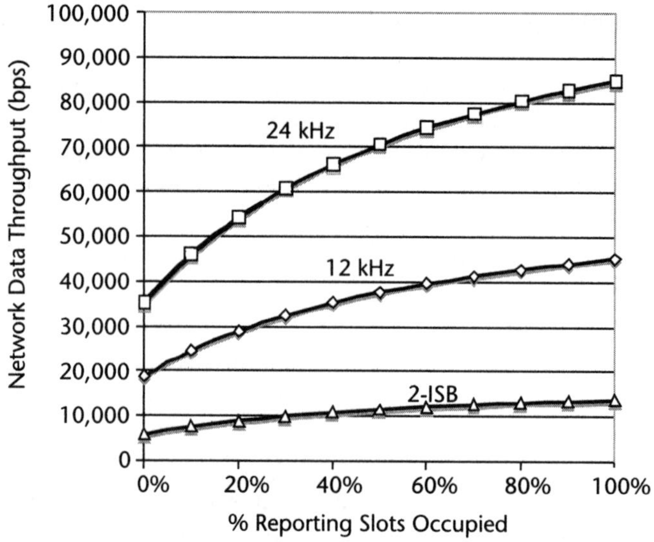

Figure 6.10 Peer-to-peer throughput in COP scenario.

polling case, each complete polling cycle requires ten link turnarounds as Node A sends the token to each other station. That station then returns the token, along with any ACKs and data. In the peer-to-peer case, a complete token rotation requires only six link turnarounds. Each node sends data, ACKs, and then the token.

6.5.4 Robust Voice Communications

Yet another intriguing new capability offered by the WBHF waveforms is the ability to send digital voice beyond line-of-sight at very low SNR. The recently standardized 600 bps MELP vocoder (STANAG 4591) [16] offers good voice intelligibility at BER up to about 10^{-2}. Using the very robust Walsh-coded 600-bps waveform for 24 kHz channels, we should be able to communicate voice reliably at signal levels well below the noise floor.

6.6 On-Air Testing

Development of engineering prototypes of WBHF modems began even before the standardization process was initiated. As a result, a substantial body of experience with these waveforms over the air was available while the standard was being completed. Some of these results are summarized here.

6.6.1 Harris On-Air Testing

In April 2011, on-air testing of the new wideband HF data modem standard was performed on a Harris-developed prototype implementation of US MIL-STD-188-110C. The goals of the on air testing were threefold: (1) test and evaluate the operation of the prototype wideband HF data modem; (2) gain experience operating a wideband HF system over a short range, NVIS link, that is typical of tactical military scenarios; and (3) perform initial on-air test and evaluation of the Harris Spectrum sensing approach as a component of a future wideband HF automated link establishment system (see Chapter 7).

6.6.1.1 Equipment and Link Details

The link under test is a predominantly east-west link between the Harris RF Communications Division in Rochester, NY, and a leased test facility located in the Stockbridge, NY, area. The distance between sites is approximately 100 miles. The Rochester site is an urban location with a fairly high noise environment. The Stockbridge site is a rural location with a low noise environment.

The Rochester site utilized a Harris RF-5800 based 400-watt HF radio system and a wideband dipole antenna. The Verona-Stockbridge site utilized a Harris RF-5800 125-watt system including a prepost selector, a RF-382 antenna coupler and a RF-1912 fan dipole antenna. Both sites also included a prototype wideband HF system and a laptop computer for control and data logging. In keeping with the goal of testing a tactical HF communications link, both antennas—while not mobile whip antennas—could be considered field-expedient antennas that could be set up relatively quickly. Testing was performed over several days from approximately 8:00 AM to 5:00 PM local time.

Figure 6.11 contains the VOACAP prediction of propagation over this link. Accurate antenna models were utilized, and the link was evaluated for an average power of 200 watts. This prediction proved to be extremely accurate in terms of

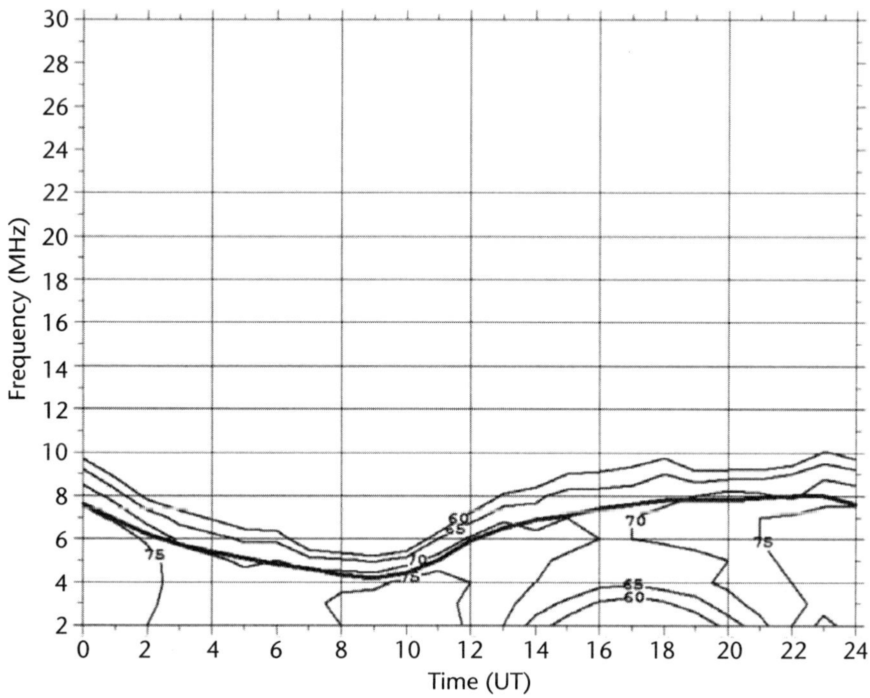

Figure 6.11 VOACAP propagation prediction from Rochester to Stockbridge, NY. (After [17].)

the propagating frequency range. All daytime data transfers were done on allocated frequencies in the 4- to 7-MHz range.

6.6.1.2 Test Procedure

The following procedure was used throughout the wideband HF test:

- Each test began with a STANAG 4538 3G ALE LQA exchange originating at the Rochester site. This provided a mechanism to determine which frequencies in the frequency set were propagating, and their SNRs and multipath spreads. (These frequencies were allocated by the FCC under a special temporary authorization for wideband HF test and evaluation.) The 3G ALE LQA exchange only utilized the 3 kHz just above the center frequency.
- Several of these frequencies were selected, and a spectrum sense was run on each of the selected channels. The received signal strength was also noted for each of these frequencies.
- Finally, the test frequency was selected, along with a bandwidth and waveform type. It should be noted that, although the goal was to send data over the link, the best or optimum frequency was not always selected. Frequencies with varying amounts of multipath and fading were selected, based on 3G ALE link estimates, to test the utility of wideband HF over various channel conditions.

During each wideband HF data modem reception, bit errors, 1000-bit packets containing errors, estimated SNR and the HF channel impulse response were logged by the computer. This test procedure was then repeated throughout the day.

6.6.1.3 Test Results

Table 6.17 summarizes the results of several days of testing over the Rochester-Stockbridge link. For each bandwidth and bit rate combination tested, the table displays the total number of on-air seconds, summed over all tests attempted. The table also illustrates how many seconds were error-free and the amount of error-free data delivered, in MB, throughout the duration of the on-air tests. The last row in the table shows that there were a total of 20,367 on-air seconds, 18,033 of which were error-free. During the course of the on-air test, almost 128 MB of data was transferred from Rochester to Stockbridge. The table also shows that significant portions of the test time and data transferred were at bit rates of 51,200, 64,000, and 76,800 bps.

Figures 6.12 and 6.13 display SNR and error profiles collected approximately between 12:00 and 13:00 on two adjacent days, conveniently automatically measured during the lunch break. Figure 6.12 shows a clear correlation, as expected, between deep SNR fades and the occurrence of errors. Figure 6.13 experienced a single burst of errors during a period of lower SNR accompanied by fast fading.

Figures 6.14(a) and (b) display typical HF channel impulse responses as measured by the wideband HF data modem during the test. The channel impulse response is an estimate of the amount of received signal power versus time delay,

Table 6.17 Summary of Test Results, Rochester to Stockbridge, NY, April 2011

Bandwidth (kHz)	*Bit Rate (bps)*	*Total Seconds*	*Error-Free Seconds*	*Percent Error-Free*	*Error-Free Data Transferred (MB)*
12	32,000	826	826	100	3.3
12	38,400	720	662	92	3.2
21	38,400	620	620	100	3.0
18	48,000	586	566	96.6	3.4
24	51,200	6100	6007	98.5	38.4
24	64,000	9596	8272	86.2	66.2
24	76,800	1919	1080	56	10.4
Total		20367	18033	88.5	127.9

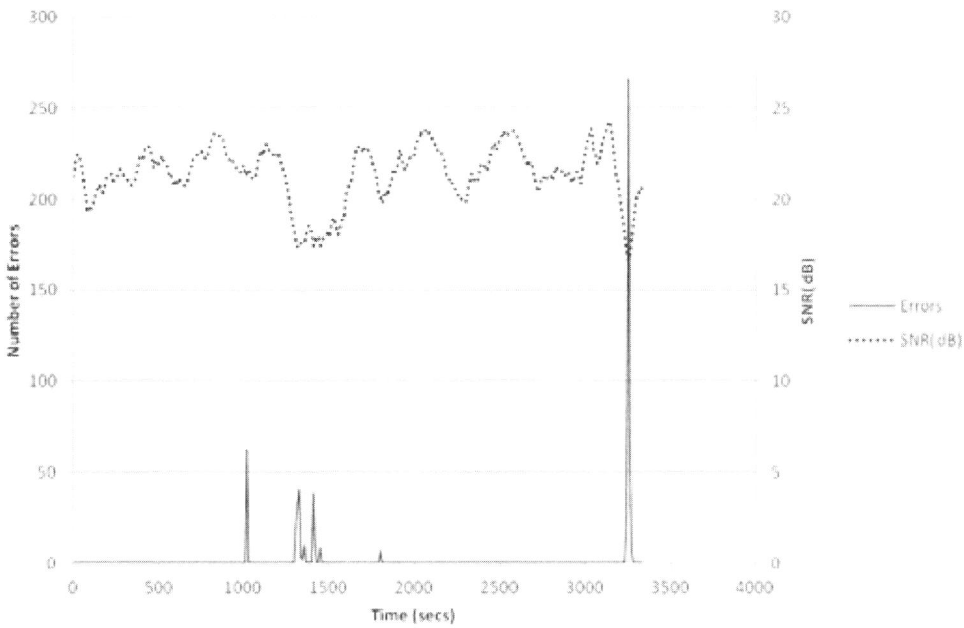

Figure 6.12 24 kHz, 64 kbps, 95.4% error-free, 25.45 MB error-free. (Reprinted with permission from [17].)

and gives an immediate measure of the number of propagating modes or paths, the relative time delay between them, and their relative power. This estimate is continuously updated throughout the reception as the HF data modem tracks the changing HF channel conditions.

Although these snapshots come from two different tests, they are very similar. Both show the presence of three modes or paths. The first two paths are relatively close in delay, 0.15 ms in Figure 6.14(a) and 0.3 ms in Figure 6.14(b). Both cases show a third attenuated delayed path at approximately 2 ms from the initial path. It should be noted that the delays of interest are the relative delays between the paths and that there is no significance attached to the absolute delay, which is merely where the data modem sets its alignment to the received signal. Also note that these paths are dynamic and fading during the reception; in fact the spectrum

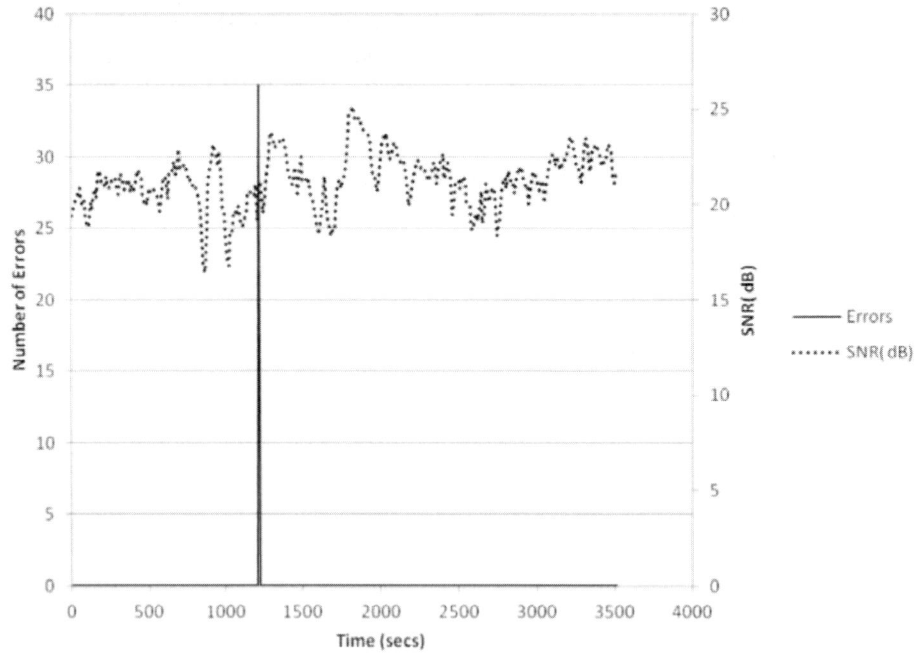

Figure 6.13 24 kHz, 51.2 kbps, 99.7% error-free, 22.45 MB error-free. (Reprinted with permission from [17].)

of this fading process is directly related to the definition of Doppler spread or fade rate.

6.6.1.4 Long-Haul Tests

Testing similar to that described in the last section was repeated over a link from Rochester, NY, to Melbourne, FL. The equipment was identical to the previous test; however the antennas were changed to log-periodic antennas at both ends of the link.

Table 6.18 highlights the results of three days of this test. A total of 324.76 MB of data was transferred. Throughout this test 84% of the received seconds of data were received error free.

6.6.2 Rockwell-Collins On-Air Testing

Independent development of prototype wideband HF modems was underway at both Rockwell Collins and Harris Corporation when Eric Johnson, chair of the MIL-STD-188-110 Technical Advisory Committee (TAC), became aware of the parallel developments. He approached both companies with the suggestion that it would be better for all concerned if, rather than developing competing products, an interoperable standard could be produced.

Invitations were sent to both companies, as well as other TAC members, and on August 5, 2009 the first wideband HF workshop was held at the Physical Science Laboratory at New Mexico State University, Las Cruces, NM. At the meeting, it became clear that Harris and Rockwell Collins were pursuing similar approaches

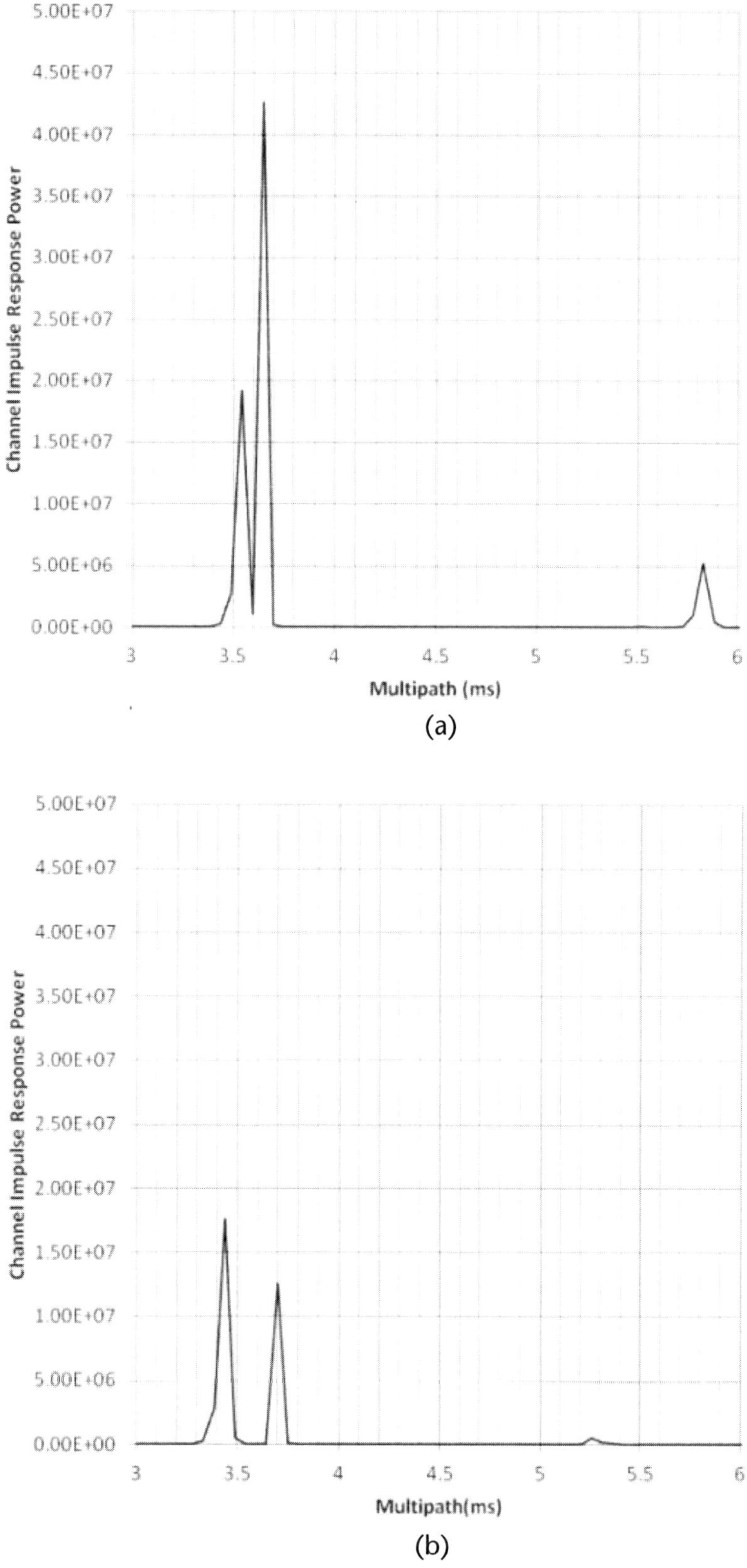

Figure 6.14 HF channel impulse response. (Reprinted with permission from [17].)

Table 6.18 Data Transferred Rochester-Melbourne

Bandwidth (kHz)	*Bit Rate*	*Total Seconds*	*Error-Free Seconds*	*Error-Free %*	*Error-Free MB*
24	12,800	622	622	100	0.995
3	16,000	51	38	75	0.076
6	24,000	1179	854	72.4	2.55
24	25,600	296	296	100	0.94
24	38,400	660	611	92.6	2.94
12	48,000	595	489	82.2	2.93
24	51,200	5904	4794	81.2	29.488
18	57,600	261	95	36.4	0.69
24	64,000	12,525	11,597	92.6	95.32
24	76,800	9583	7330	76.5	70.366
24	96,000	11,874	9868	83.1	118.46

to the wideband HF problem. Both were developing serial-tone approaches, with QAM constellations for the higher data rates, and it became clear at the meeting that an interoperable standard could be achieved with relatively modest changes to both design approaches.

From the Rockwell Collins perspective, the most significant change as a result of the workshop was the decision to look at bandwidths up to 24 kHz, where the design under development had assumed only 12 kHz would be available. The outcome of the meeting was a decision to pursue a collaborative development of a wideband HF standard, with a second workshop to be held in November, and a draft to be presented to the TAC for consideration in early 2010.

The wideband HF modem prototype under development at Rockwell Collins used waveforms very similar to those now found in Appendix D, for 6- and 12-kHz bandwidths.

- The waveform used the same 256QAM and 64QAM constellations, as well as having the same block sizes for unknown data and known probe sequences at the highest data rates.
- The preamble was quite different, being very similar in nature to the STANAG 4539 preamble, rather than using the Walsh symbol approach of the MIL-STD serial tone. For the higher data rates, this approach provided more than adequate acquisition characteristics.

While it was clear that the final MIL-STD wideband waveform design would differ from the initial prototype, the decision was made to complete the development and to conduct on-air testing with it in order to develop a better appreciation of any issues that might arise in the fielding of wideband waveforms. Because of the high degree of commonality anticipated between the prototype and the final design, it seemed likely that performance of the final design, at least for bandwidths up to 12 kHz, could be extrapolated from results obtained with the prototype. In the fall of 2009, the implementation was complete, and FCC approval for on-air testing of the prototype was obtained.

The first on-air testing of the 12-kHz prototype wideband HF waveform began in January of 2010, with local tests near Cedar Rapids, IA, using surface-wave propagation. Data rates of up to 32 kbps were achieved.

On February 12, skywave tests conducted between Cedar Rapids, IA, and Richardson, TX, achieved data rates of 38.4 kbps. On February 17, diversity reception was used at Richardson, TX, to allow successful operation with data rates up to 64 kbps. As indicated previously, for the 12 kHz bandwidth this data rate was not expected to be viable for sky-wave operation, so this early success was a pleasant surprise. All tests used good antennas: rotatable log-periodic, hex log-periodic, and large omnidirectional monopoles, with transmit power varying from 90W to 4 kW. In many cases, the higher data rates were successfully obtained with transmit powers of 200 W or less.

Figures 6.15 and 6.16 show constellations for received 12-kHz waveforms [18]. In Figure 6.15, we see reception of 38.4 kbps without diversity, while Figure 6.16 shows reception of 64 kbps with diversity.

Figure 6.17 shows the block errors that were observed during an extended diversity reception at 64 kbps. Each block contains 1000 bits, and each increment on the horizontal axis of the figure contains 1000 blocks. A single bit in error in a block results in a block error. An examination of the figure shows many error free intervals. The most highly errored segment still has more than 80% error free blocks. Clearly this would support data transfer with an ARQ scheme very effectively.

The next major step in WBHF testing and development at Rockwell Collins was participation in the March 2011 Trident Warrior exercise. For this exercise, prototype WBHF systems were installed at four sites in North America (Figure 6.18): Ottawa (Ontario, Canada); Cedar Rapids, IA; Richardson, TX; and Las Cruces, NM. Path lengths ranged from roughly 1000 km to 3000 km.

Figure 6.15 Received constellation—38.4 kbps in 12 kHz, no diversity. (© 2010 IEEE. Reprinted with permission from [18].)

Figure 6.16 Received constellation—64 kbps in 12 kHz, with diversity. (© 2010 IEEE. Reprinted with permission from [18].)

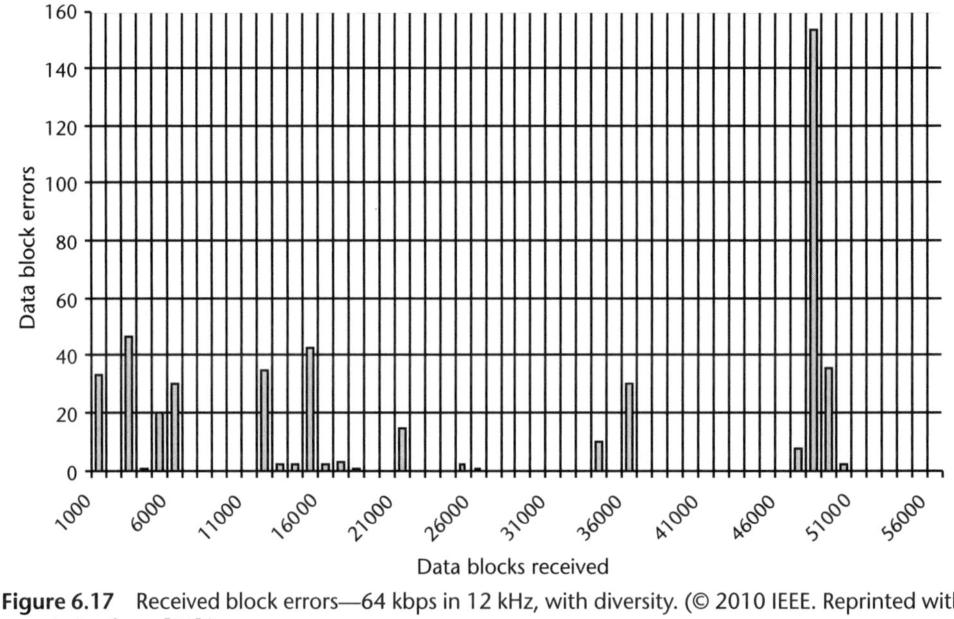

Figure 6.17 Received block errors—64 kbps in 12 kHz, with diversity. (© 2010 IEEE. Reprinted with permission from [18].)

- The modems and RF equipment were provided by Rockwell-Collins. Cedar Rapids used a 4 kW system, while the other sites were limited to 1 kW. All systems were limited to 18-kHz bandwidth during the Trident Warrior exercise, but 24-kHz bandwidths were available for later testing.
- Networking controllers implementing the STANAG 5066 token-passing protocol (see Chapter 3, Section 3.4) were provided by the U.S. Navy SPAWAR Systems Center in San Diego, CA.

Figure 6.18 Trident Warrior on-air testing.

- Las Cruces and Cedar Rapids used rotatable log-periodic antennas. Richardson used an omnidirectional TCI-CMV330 low takeoff angle HF antenna, and Ottawa used a sloping V antenna.

6.6.2.1 WBHF Application Demonstrations

The ability of WBHF to deliver real-time video over skywave paths was amply demonstrated using this prototype WBHF network. In one test, full-motion, color H.264 video was streamed at 38.4 kbps (18-kHz bandwidth) from Las Cruces to Cedar Rapids (1700 km) for 75 minutes without sync loss. The video stream was 15 frames per second, with frame size scaled to data rate. Data rates demonstrated ranged from 19.2 to 120 kbps.

Reliable file transfer over HF skywave channels was also demonstrated using this network, including the successful delivery of an FTP server program (the FileZilla Server installer, a 1.6 MB file) from Las Cruces to Ottawa via WBHF. This file executed successfully following its reception and installation. Several file types were subsequently exchanged between the two sites using that FTP server application. This was almost certainly the first instance of open-source software distribution via wideband HF!

6.6.2.2 Wideband HFIP Demonstration over Skywave

The U.S. Navy HFIP system is an implementation of the Internet protocol and Internet applications using the STANAG 5066 HF subnetwork service, using the Annex L token-passing MAC protocol. It was designed to operate within a battle group using a shared (surface-wave) channel. The Trident Warrior HFIP test explored issues involved in using this system over skywave links.

With the wide range of path lengths and orientations among the test stations, it was challenging to find a single frequency that propagated well on all links. Nevertheless, a four-node token-passing network was created on a compromise frequency. Link speeds varied from 19.2 to 38.4 kbps (in an 18-kHz channel).

Use of HFIP in a skywave network can benefit from split-frequency operation, wherein different frequencies are used for transmitting and receiving. For example, the network shown in Figure 6.19 could operate on separately selected frequencies, as shown in Figure 6.16. (These frequencies were selected using VOACAP for the conditions prevailing during the test.)

However, for full operation, each station would need to listen for traffic, acknowledgments, and the token on both frequencies shown. A single transceiver could be used if the MAC protocol kept track of which station held the token, and maintained a table of agreed frequencies that each station would use for transmitting to each other station.

6.6.2.3 Skywave Throughput Measurements

During the Trident Warrior exercise, the HFIP protocol was used to set up an IP network over the HF communications links established with WBHF. One of the objectives of the exercise was establishing what user data rates could be supported over the network for various modem and HFIP settings. Table 6.19 shows representative user throughputs achieved on links among the four sites in various combinations. The impact of the TCP window size clearly demonstrates the interaction between the MAC layer protocol and the transport layer protocol in this environment. For

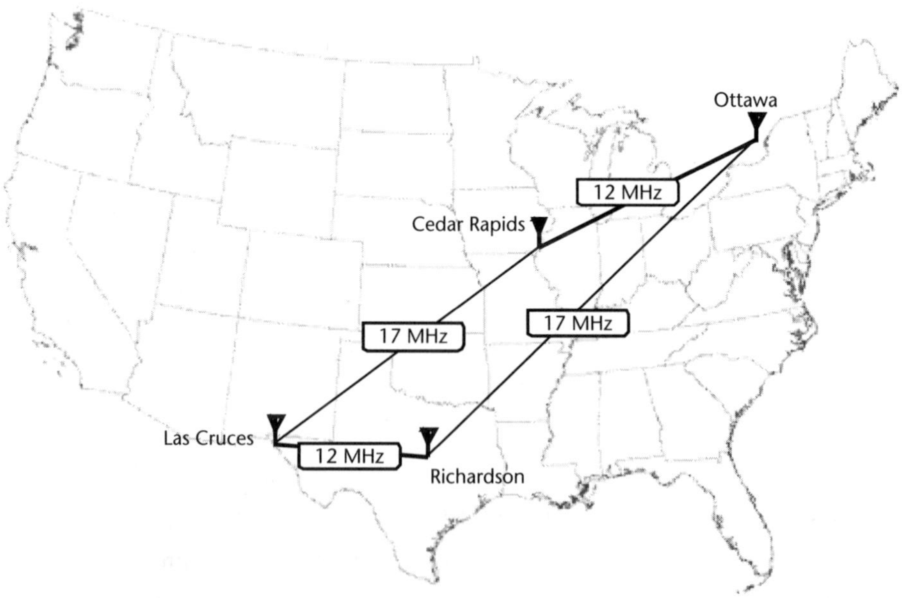

Figure 6.19 Multifrequency HFIP ring.

Table 6.19 Trident Warrior Skywave Throughput Measurements

Nodes	*Modem Data Rate*	*Interleaver*	*TCP Window*	*Throughput*
Cedar Rapids and Richardson	38.4	Short	64kB	24.1 kbps
	57.6	Short	64kB	22.6 kbps
	57.6	Short	64kB	35.8 kbps
Las Cruces and Cedar Rapids	57.6	Ultrashort	64kB	14.5 kbps
	57.6	Ultrashort	64kB	14.0 kbps
	57.6	Ultrashort	500kB	46.9 kbps
Cedar Rapids (CR) and Ottawa (O)	72 (CR), 48 (O)	Short	64kB	17.3 kbps
	72 (CR), 48 (O)	Short	500kB	53.5 kbps

this reason, when TCP is employed over links of this type, a TCP proxy is highly desirable.

6.6.2.4 WBHF Interleaver Performance Study

One factor that engineers consider when implementing operational HF networks is the interleaver size. In simulation, longer interleavers almost always provide significantly better performance as measured by bit error rates for a given signal-to-noise ratio. The trade-off that has to be made in the design of a system is deciding whether the latency that longer interleavers impose can be tolerated by the particular application. For token passing schemes (such as HF-IP) or TDMA protocols (such as MARLIN, Draft STANAG 4691), the additional latency translates into longer cycle times, and typically, much poorer user experience from the point of view of network responsiveness. As a result, designers of these kinds of systems usually use one of the two shortest interleavers for these systems. There has been some limited over-the-air testing conducted with the objective of determining whether the performance penalties seen in simulation are reflective of real operational conditions.

Figures 6.20 through 6.23 show the result of such a test [19]. In this case, transmission originates in Cedar Rapids, IA, on a log-periodic antenna, with a 250 W-average power, and is received in Las Cruces, NM, by a log periodic antenna. The four examples all use the 76.8-kbps waveform in a 24-kHz channel, but with each of the four interleavers represented. Recall that each successive interleaver selection differs by a factor of four from the adjacent setting, and the latency penalty scales in the same fashion. In this data set, it is apparent that the long interleaver (approximately 8 s) offers more error free intervals and generally better performance. The relative merits of the other three interleavers are not so clear cut, with the ultra-short appearing to be better than the short interleaver, for example.

From this limited testing (mostly obtained on a single mid-latitude, medium-range HF path), it would appear that the long interleaver consistently outperforms the shorter interleaver settings, but there is often very little difference (on average) among the results obtained with those shorter interleavers. Further research over a much wider range of paths is clearly required. However, contrary to popular wisdom (at least in this case), choosing the interleaver setting that best fits the application may not impose a significant performance penalty.

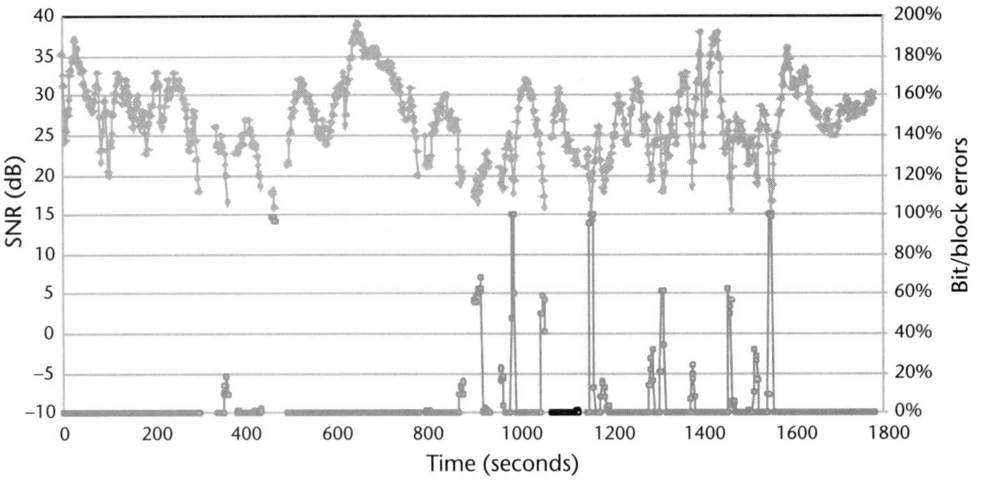

Figure 6.20 24 kHz BW, 76.8 kbps, long interleaver. (Source: [19].)

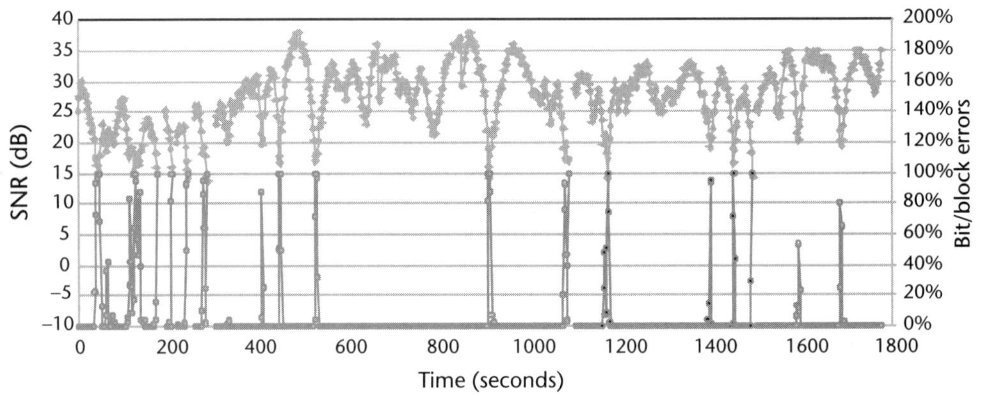

Figure 6.21 24 kHz BW, 76.8 kbps, medium interleaver. (Source: [19].)

6.7 Operational Considerations

To conclude this discussion of WBHF capabilities, we now address some of the operational considerations that arise in using the new waveforms.

First, a WBHF system provides waveforms that offer the same data rate in different bandwidths. As a result, we can trade off extra bandwidth to get more robustness at the same data rate and transmit power. For example, consider the 9600-bps waveforms available in 3, 6, and 24 kHz.

If we spread a fixed total power over a 6 kHz channel rather than 3 kHz, our SNR will drop by 3 dB; this is a good trade-off when we see that the modulation required to send data at 9600 bps drops from 64-QAM to 8 PSK, with the result that 12 dB less SNR will be required for a BER of 1E-5 (Table 6.20). (This ignores the added benefit of a reduced PAR with the simpler waveform.) A further improvement is available if a 24-kHz channel is available: the constant-power SNR

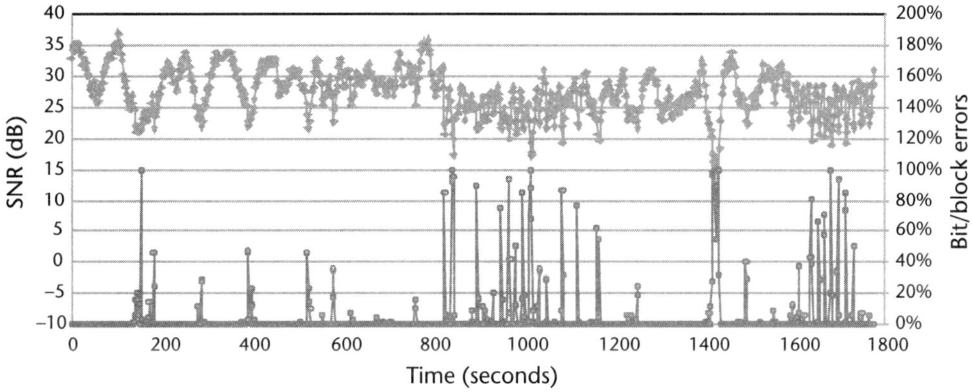

Figure 6.22 24 kHz BW, 76.8 kbps, short interleaver. (Source: [19].)

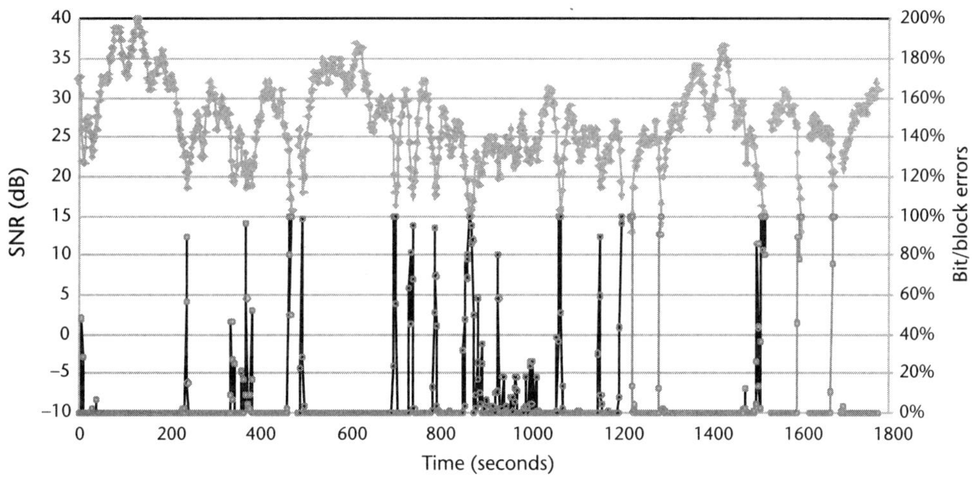

Figure 6.23 24 kHz BW, 76.8 kbps, ultrashort interleaver. (Source: [19].)

Table 6.20 9600 bps Waveform Trade-Off

Bandwidth	*9600 bps WID*	*9600 bps Modulation*	*SNR for 1E-5 BER*
3	10	64-QAM	31
6	7	8-PSK	19
24	4	2-PSK	10

loss in expanding from 6 kHz to 24 kHz is 6 dB, while a 9 dB improvement in SNR robustness should be available.

An operational drawback with the WBHF waveforms is the absence of the reinserted preambles that were present in the 3200 to 9600 bps narrowband waveforms, and in the Rockwell Collins WBHF prototype. These periodically announced the data rate and interleaver in use for a transmission, and allowed receivers that missed the initial synchronization preamble to autobaud on data. This can be useful in broadcast applications, but is difficult to implement consistently across eight different symbol rates corresponding to the eight bandwidth selections. During the development of the WBHF waveform, discussions with the user community seemed to indicate that, on those occasions where there was a need for a sync-on-data capability (e.g., a broadcast transmission), it was reasonable to assume that the parameters of the transmission would be known in advance. As a result, the WBHF waveform includes the capability to sync-on-data, but only if the data rate, interleaver, and constraint length are known to the receiver in advance.

A similar lack is the absence of any autobandwidth information in the waveform. The thinking that led to this choice was the belief that most systems would either be fixed bandwidth, or would require some kind of ALE function to determine the available bandwidth prior to transmission of the WBHF waveform. The design of the ALE was considered to be outside of the purview of the 110C modem specification.

So far we have assumed that WBHF channel allocations will be usable as assigned (i.e., that all of a 24 kHz allocation will be usable). However, this ideal case may not always be realizable in practice due to interference on portions of the allocation. What would the impact of such partial-band interference be? The overall SNR would suffer, and the data rate would need to be reduced below what might be achievable if we could identify and use the clear portion of the WBHF channel. This latter capability has not yet been standardized, but is an active area of research and will be discussed in the next chapter.

References

[1] Brakemeier, A., "Criteria to Select Proper Modulation Schemes," *Nordic HF-95 Conference*, August 15–17 Fårö, Sweden: 1995, Section 3.3.1.

[2] Johnson, E., "Performance Envelope of Broadband HF Data Waveforms," *Proceedings of MILCOM 2009, IEEE*, Boston, MA: 2009.

[3] Johnson, E., "Simulation Results for Third Generation HF Automatic Link Establishment," *Proceedings of MILCOM '99, IEEE*, Atlantic City, NJ: 1999.

[4] Johnson, E., M. Balakrishnan, and Z. Tang, "Impact of Turnaround Time on Wireless MAC Protocols," *Proceedings of MILCOM 2003, IEEE*, Boston, MA: 2003.

[5] Martone, M., *Multi-Antenna Digital Radio Transmission*, Norwood, MA: Artech House, 2002.

[6] Sinha, N. B., R. Bera, and M. Mitra, "Capacity and V-BLAST Techniques for MIMO Wireless Channel," *Journal of Theoretical and Applied Information Technology*, Vol. 4, No. 1, 2005

[7] Holter, B., "On the Capacity of the MIMO Channel—A Tutorial Introduction," Norwegian University of Science and Technology, http://new.iet.ntnu.no/projects/beats/Documents/MIMO_introduction.pdf.

[8] Jorgenson, M., et al., "The Evolution of a 64 kbps HF Data Modem," *IEE Eight International Conference on HF Radio Systems and Techniques*, University of Surrey, Guildford, UK, July 2000.

[9] Recommendation ITU-R F.1487, "Testing of HF Modems with Bandwidths of up to about 12 kHz using Ionospheric Channel Simulators," *International Telecommunication Union*, Geneva, Switzerland: 2000.

[10] Elvy, S., "High Data Rate Communications over HF Channels," *Nordic HF 98 Conference Proceedings*, Fårö, Sweden: 1998.

[11] ITU, "Recommendation 520-1 Use of High Frequency Ionospheric Channel Simulators," *Recommendations and Reports of the CCIR*, Vol. III, Geneva, Switzerland, 1982, pp. 57–58.

[12] STANAG 4415, "Characteristics of a Robust, Non Hopping, Serial-Tone Modulator/Demodulator for Severely Degraded HF Radio Links," *North Atlantic Treaty Organization*, Edition 1, December 24, 1997.

[13] MIL-STD-188-110B, "Military Standard - Interoperability and Performance Standards for Data Modems," *United States Department of Defense*, May 27, 2000. (Current version is MIL-STD-188-110C, dated Septamber 12, 2011.)

[14] Yasuda, Y., K. Kashiki, and Y. Hirata, "High-Rate Punctured Convolutional Codes for Soft Decision Viterbi Decoding," *IEEE Transactions on Communications*, Vol. COM-32, No. 3, March 1984.

[15] Maslin, N., *HF Communications: A Systems Approach*, London, U.K.: Plenum Press, 1987.

[16] STANAG 4591, "The 600 Bit/s, 1200 Bit/s, and 2400 Bit/s NATO Interoperable Narrow Band Voice Coder," *North Atlantic Treaty Organization*, Edition 1, October 3, 2008.

[17] Furman, W. N., and J. W. Nieto, "Recent On-Air testing of the New Wideband HF Data Modem Standard, U.S. MIL-STD-188-110C," *Proceedings of IES 2011, the 13th International Ionospheric Effects Symposium*, Alexandria, VA, May 2011. Available at www.NTIS.gov.

[18] Jorgenson, M., et al., "Implementation and On-Air Testing of a 64 kbps Wideband HF Data Waveform," *Proceedings of MILCOM 2010, IEEE*, San Jose, CA: 2010.

[19] Jorgenson, M., et al., "WBHF Skywave Interleaver Performance Test Results," *HF Industry Association Meeting*, January 2012, San Diego, CA, 2012. Available at http://www.hfindustry.com.

CHAPTER 7

Future Directions

In this final chapter, we look ahead to promising new technologies either envisioned or under development for HF radio, starting with the need for automatic link establishment (ALE) for wideband HF data applications.

7.1 Wideband ALE

The new wideband HF (WBHF) data waveforms described in Chapter 6 offer an exciting new capability to transfer data over HF channels at speeds up to 120 kbps. Waveforms are specified for bandwidths from 3 through 24 kHz, in increments of 3 kHz. Why would we want waveforms for such "odd" bandwidths as 9 kHz or 21 kHz? The waveform designers anticipated that, in real-world HF applications, we may have a 12- or 24-kHz allocation, but parts of that wide channel might experience interference. This would render the channel unusable unless we could contract our active bandwidth to match the clear subchannel within the allocated channel.

In narrowband HF applications, we automated the process of finding a usable frequency and setting up links between two or more radios. This is termed automatic link establishment (ALE). With our new bandwidth flexibility, we now need additional automated capabilities: (1) detect and characterize interference within a wideband channel, and (2) coordinate use of the clear subchannel.

This new spectrum management capability is variously termed *wideband ALE* (WBALE) or *fourth generation ALE* (4G ALE). In this chapter, we discuss requirements and design objectives for such a technique, explore some candidate design concepts, and present early results from on-air testing aimed at verifying the suitability and feasibility of these concepts [1].

7.1.1 Wideband ALE Design Considerations

The new wideband HF waveforms offer a variety of modulations, FEC code rates and constraint lengths, and interleaver depths. In adaptive narrowband HF systems, selecting an operating point among these options usually falls to the ARQ process, while ALE is engaged only to change frequency. In a wideband system, an integrated adaptive control process (i.e., the WBALE process) might manage all of the preceding options, as well as determine the optimal bandwidth and frequency offset within an allocated channel. Ideally, the WBALE process will apply some intelligence to achieve the best possible communications performance and quality of service using the infamously time-varying and unpredictable HF channel.

The literature [2, 3] reports the potential application of cognitive radio techniques in the ALE function of an HF communications system. The design of a cognitive wideband ALE function for wideband HF will have to address a number of technical challenges:

- Channel bandwidth is a new variable to be considered in allocating, managing, and selecting frequencies. WBHF users will want channels wider than 3 kHz, up to and including 24 kHz. Because the spectrum available for assignment to any network is limited, assigned channels may have a mixture of bandwidths.
- In selecting the signal constellation, code rate and constraint length, and interleaving to be used in a WBHF transmission, a WBALE system will need to estimate the propagation characteristics of the available channels. Our initial work assumes that HF channels (including skywave channels) of up to 24 kHz exhibit sufficiently uniform Doppler and multipath characteristics so that a 3-kHz probe waveform can be used to adequately characterize the wider channel [2].
- In using wider bandwidths, a WBHF communications system becomes inherently more vulnerable to interference: a wider-bandwidth channel represents a larger cross section to interferers. An allocation doesn't guarantee interference-free availability of a channel. Even if a nation's regulatory authorities intend to provide exclusive allocations to specific users, frequency reuse often occurs across national borders and land/sea boundaries. Frequency reuse is becoming an accepted feature of HF frequency management as regulators try to maximize the communications capacity that can be provided within spectrum limits [4].

Narrowband interference (e.g., by users of 3-kHz channels) within wideband channels cannot be ruled out, especially in the presence of frequency reuse. Users of the new WBHF waveforms need some form of bandwidth agility so that interference from a 3-kHz transmission will not block use of an entire 24-kHz WBHF channel. The various bandwidths in the waveform family make it possible for a WBHF system to use the remaining portion of the wideband channel that is not occupied by the interfering signal.

A WBALE system capable of meeting these challenges will need a spectrum sensing capability that is able to listen on an entire wideband channel of 24 kHz (or more), detect and evaluate any interfering signals on the channel, and identify a portion of the channel that may be usable. The reliability and accuracy of this spectrum sensing function will usually be a significant factor determining the performance of a wideband HF system.

7.1.2 A Conceptual WBALE System

We can enhance current 3G equipment for use in WBALE experiments. For the experimental system to be evaluated in this chapter, we assume (1) coexistence with fast link setup (FLSU); (2) spectrum sensing, as described above; (3) support for IP-over-HF.

7.1.2.1 FLSU Coexistence

WBALE will need to coexist with STANAG 4538 FLSU, so that a WBHF-capable station can use WBHF technology when linked with other WBHF-capable stations, while also participating in FLSU networks. This leads to the following design assumptions:

- The WBALE scanning procedure will be the same as STANAG 4538 FLSU, but with a dwell time of 1.35 seconds per channel [5].
- WBALE link establishment and control signaling will use 3-kHz burst waveforms similar to those of STANAG 4538 FLSU.
- WBALE link setup will begin with transmission of a request PDU similar to an FLSU_Request PDU, and using the same acquisition preamble and timing. This enables WBALE receivers to search simultaneously for both FLSU and WBALE calls.
- Each wideband channel will include a 3-kHz subchannel that is shared with the FLSU network. This subchannel will be the portion of each wideband channel used for link setup by WBALE.

7.1.2.2 Spectrum Sensing

WBALE stations will be able to detect occupancy or interference within any portion of a channel allocation of up to 24 kHz. This spectrum sensing capability is assumed to operate as follows:

- WBALE stations will sense the state of the entire wideband channel during each dwell period as they scan.
- The dwell period is too short for a thorough characterization of channel occupancy and interference during a single dwell. Therefore, stations will maintain a time-weighted average of spectrum snapshots from recent dwell periods for each channel.
- Some types of interference are present only in short bursts (e.g., the 3G burst waveforms). To avoid colliding with such signals, the WBALE system will collect additional spectral occupancy statistics, such as the maximum recently-observed energy level in each portion of the band being sensed.

Note that interference avoidance using spectrum sensing is vulnerable to the hidden terminal problem familiar from other applications of carrier sensing: when station A wants to send traffic to station B, station A may not be able to detect a distant transmission that would interfere at station B. This may require the use of a protocol for stations to share measurements of their local spectral environments. However, such a protocol is not further discussed here.

7.1.2.3 Internet Support

We also assume that WBALE will be designed for efficient support of IP-over-HF service, because many applications that need the throughput of wideband HF are IP-based. This suggests the following assumptions:

- Latency in data delivery and in link setup and maintenance must be minimized.
- Data transfer mechanisms for wideband HF will be designed to minimize the need for time-consuming logical link turnarounds (see Chapter 5, Section 5.8.1.1). For example, in the 3G HDL and LDL protocols, payloads can flow in only one direction at a time. This delays traffic in one direction on a point-to-point link while traffic in the other direction is being delivered.
- Differentiated service will be available to traffic in distinct service classes, to support the Diffserv facility of RFC 2474 [6] or other quality-of-service management mechanisms.

7.1.3 Spectrum "Sense and Avoid" Demonstrated

Harris RF Communications implemented a spectrum sensing capability in the prototype radios used for WBHF field testing (see Chapter 6, Section 6.6.1) [7]. After the radio measures energy levels over a 24-kHz band of interest, the measurements are formatted and displayed on a laptop computer. For example, Figure 7.1 shows an entire channel free of interference; only the noise floor is visible. Figure 7.2 shows a 3-kHz serial tone modem probe signal in the center of the channel.

The WBALE techniques of spectrum sensing and subchanneling to avoid interference are illustrated using observations from on-air testing of the wideband waveforms on an NVIS path from Rochester, NY, to Verona, NY. Figure 7.3 shows an example of the interference that could be encountered when using WBHF in real-world applications. The frequency shown here was officially assigned to the Harris Corporation for this experiment. However, an attempt to transfer data at 64

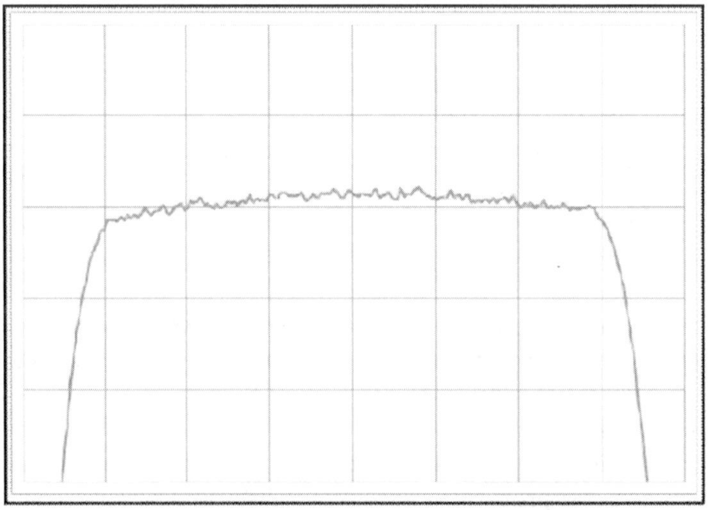

Figure 7.1 A vacant 24-kHz channel. (© 2012 IET. Reprinted with permission from [1].)

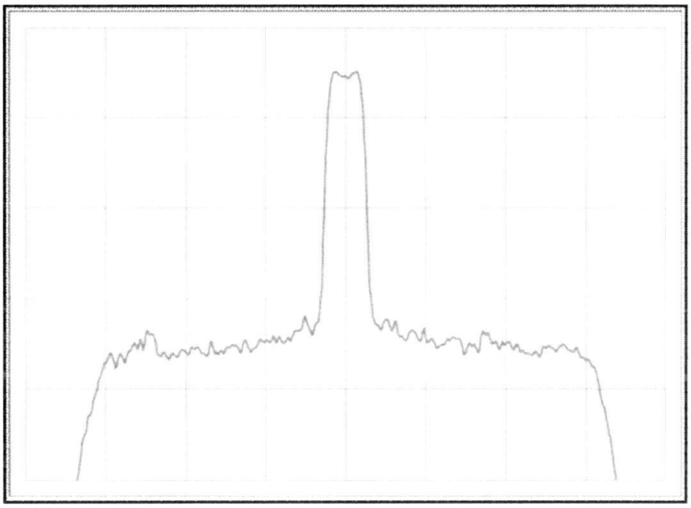

Figure 7.2 A 3-kHz probe in a 24-kHz channel. (© 2012 IET. Reprinted with permission from [1].)

kbps with a 24-kHz bandwidth would have failed completely (50% BER). The reason is apparent in Figure 7.3: a 1-kW AM broadcast station in Toronto, Canada, occupies the upper portion of the band.

WBALE was not implemented in this system, but by manually evaluating the spectrum to identify interference, reducing the bandwidth, and shifting the carrier frequency (as shown in Figure 7.4), data was sent error-free at 32 kbps.

7.1.4 A "Hybrid Simulation" Experiment

To estimate the potential benefits of a wideband HF data system that combines the WBHF waveforms with a cognitive WBALE, a hybrid experiment was developed.

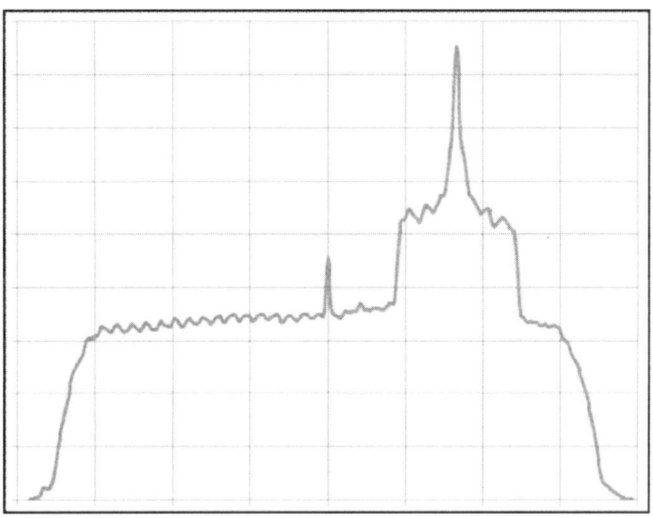

Figure 7.3 AM interference in part of a 24-kHz channel. (© 2012 IET. Reprinted with permission from [1].)

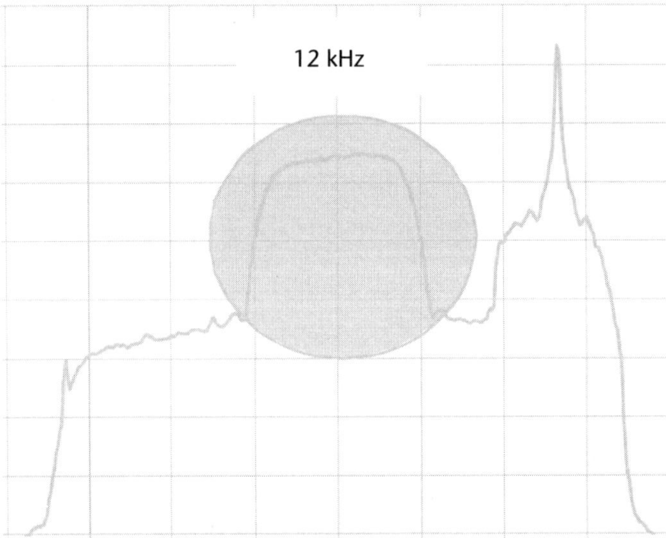

Figure 7.4 Subchannel selected to avoid interference. (© 2012 IET. Reprinted with permission from [1].)

First, VOACAP [8] was used to predict hourly ranges of usable frequencies for a link from Melbourne, FL, to Rochester, NY. The antenna configurations and transmit power levels of the Harris equipment at both sites are well known, and were used to predict received signal strengths in Rochester. Experience has shown that VOACAP predictions for this link usually are accurate.

In the on-air part of the experiment, during each hour, randomly selected frequencies within the usable frequency range identified by VOACAP were measured, one frequency per minute. The observed spectrum on each frequency was used in conjunction with the received signal strength estimate from VOACAP to predict the usable data rate of a wideband channel at that frequency as follows:

- Identify the usable bandwidth and offset for a WBHF transmission.
- Estimate the SNR in that subchannel.
- For that SNR, identify the data rate that will not exceed 10^{-5} BER (from simulation).

Aggregating the minute-by-minute data rates yielded an estimate of the total link capacity using WBHF and WBALE over the period studied.

The same procedure was used to estimate the total link capacity using the 3-kHz serial tone waveforms of Chapter 3. Comparing the two provides an estimate of the additional link capacity made possible—with no increase in transmit power—by the use of the wideband waveforms and wideband ALE.

Figure 7.5 shows how often the available bit rates were selected for both the narrowband 3-kHz modem waveforms and the wideband family of waveforms from Chapter 6. These results are for fading and multipath characteristics of the ITU-R mid-latitude disturbed (MLD) channel [9].

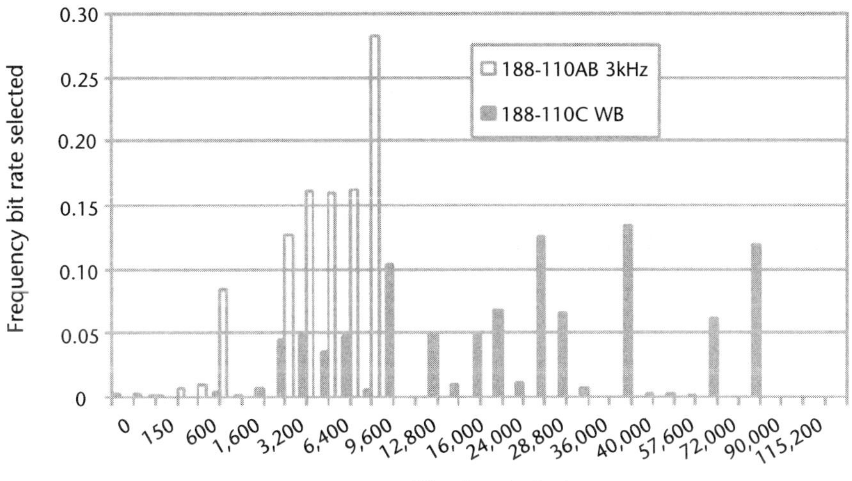

Figure 7.5 Frequency of bit rate selection. (© 2012 IET. Reprinted with permission from [1].)

The total amounts of data that the systems could have transferred in the 24 hour period simulated are presented in Table 7.1. The data capacities were calculated for both an additive white Gaussian noise (AWGN) nonfading channel and the MLD channel. Note that the increased capacity shown for the wideband system was achieved *without any increase in transmit power.*

The use of spectrum sensing played a key role in these results. More than 50% of the channels sampled had enough narrowband interference present to force the selection of a bandwidth narrower than 24 kHz. These results indicate the potential of a wideband HF system to achieve greatly increased capacity and throughput, and also the importance of using an integrated WBALE system that automates the selection of the bandwidth, subchannel alignment, and the waveform parameters to use within the subchannel.

7.1.5 WBALE Summary

The new wideband waveforms from Chapter 6 could greatly expand the range of applications that HF can support. However, with increased bandwidth come additional technical challenges that create the requirement for a new wideband ALE capability.

Initial experience with using WBHF waveforms on the air indicates that spectrum sensing will be a key element of a WBALE capability. The prototype spectrum

Table 7.1 Estimated Daily Capacity in MB

Channel	*Fixed 3 kHz*	*Adaptive Wideband*
AWGN	85 MB	505 MB
MLD	65 MB	294 MB

sensing facility reported here has been successfully demonstrated in on-air trials with the new wideband waveforms. The results of these trials suggest that WBALE systems will be able to improve significantly the throughput, capacity, and reliability of HF networks, enabling HF to serve a growing range of high-value communications applications.

The next step in developing a standardized WBALE is specifying a protocol for over-the-air negotiation of bandwidth and offset of the subchannels to use in each direction on a link, along with a data link protocol for WBHF waveforms that meets the goals for efficiency in IP applications.

7.2 Staring ALE

A more distant goal for adaptive use of the HF spectrum for long-range communications generalizes the spectrum sensing capabilities of WBALE to continuously stare at the entire HF band. This permits accumulation of continuous propagation and occupancy measurements, and eliminates the need for scanning receivers. Thus, we could theoretically achieve the short calls of a synchronized system (such as 3G ALE) without the need for synchronization.

Of course, a naive implementation of a staring receiver faces significant challenges in dynamic range. Nonetheless, conceivable advances in areas such as high-resolution analog to digital converters and tunable RF filters could vitalize this appealing concept.

References

[1] Furman, W. N., E. Koski, and J. W. Nieto, "Design Concepts for a Wideband HF ALE Capability," *IRST—Ionospheric Radio Systems and Techniques Conference*, York, UK, 2012.

[2] Furman, W. N., E. N. Koski, and J. W. Nieto, "Design and System Implications of a Family of Wideband HF Data Waveforms," *IST Symposium RTO-MP-IST-092: Military Communications and Networks, NATO Research and Technology Organisation*, Wrocław, Poland, 2010. Available at http://www.rta.nato.int/Pubs/RDP.asp?RDP=RTO-MP-IST-092, last accessed February 2012.

[3] Koski, E., and W. N. Furman, "Applying Cognitive Radio Concepts to HF Communications," *IRST—Ionospheric Radio Systems and Techniques Conference*, Edinburgh, Scotland, 2009.

[4] Arthur, N. P., I. D. Taylor, and K. D. Eddie, "Advanced HF Spectrum Management Techniques," *IRST—Ionospheric Radio Systems and Techniques Conference*, London, UK. 2006.

[5] Wadsworth, M., and E. Peach, "Initial Performance Results from an Implementation of the STANAG 4538 Fast Link Setup Protocol," *HF '01 Nordic Shortwave Conference Proceedings*, Fårö, Sweden: 2001.

[6] IETF Request For Comments RFC 2474, "Definition of the Differentiated Services Field (DS Field) in the IPv4 and IPv6 Headers," December 1998.

[7] Furman, W. N., and J. W. Nieto, "Recent On-Air testing of the New Wideband HF Data Modem Standard, U.S. MIL-STD-188-110C," *Proceedings of IES 2011, the 13th International Ionospheric Effects Symposium*, Alexandria, VA, 2011. Available at NTIS at www.NTIS.gov.

[8] Perkiömäki, J., "VOACAP Quick Guide," http://www.voacap.com/index.html, last accessed March 2012.

[9] ITU-R Recommendation F.1487, "Testing of HF Modems with Bandwidths of up to about 12 kHz using Ionospheric Channel Simulators," *International Telecommunication Union*, Geneva, Switzerland, 2000.

Acronyms and Abbreviations

1G	first generation
2G	second generation
3G	third generation
4G	fourth generation
ACK	acknowledgment
ACP	Allied Communication Publication
ACS	automatic channel selection
AGC	automatic gain control
AL	application level (in linking protection)
ALE	automatic link establishment
ALM	automatic link maintenance (protocol)
ARQ	automatic repeat request
AWGN	additive white Gaussian noise
BER	bit error ratio
BLOS	beyond line-of-sight
bps	bits per second
BW	bandwidth
COP	common operating picture
CPM	continuous phase modulation
CSMA	carrier sense multiple access
CW	continuous wave
dB	decibel
DCF	distributed coordination function (in IEEE 802.11)
DCHF	DCF modified for HF radio
DRM	Digital Radio Mondiale (international nonprofit consortium)
DSP	digital signal processor (or processing)
DTE	data terminal equipment
ELOS	extended line-of-sight
FCC	Federal Communications Commission (United States)
FEC	forward error correction

FED-STD	Federal Standard (United States)
FER	frame error ratio
FFT	fast Fourier transform
FH	Frank-Heimiller (sequence)
FPGA	field-programmable gate array
FSK	frequency shift keying
FTP	file transfer protocol
FLSU	fast link setup
HDL	high-throughput data link (protocol)
HDL+	(a higher-speed version of HDL)
HF	high frequency
HF-IP	HF internet protocol (network technology)
HFTP	HF token protocol
HMTP	HF mail transfer protocol
Hz	hertz (cycles per second)
IP	Internet protocol
ISB	independent sideband
ISI	intersymbol interference
ITU	International Telecommunications Union
JITC	Joint Interoperability Test Command (United States)
kHz	kilohertz (thousands of cycles per second)
LAN	local area network
LBT	listen before transmit
LDL	low-latency data link (protocol)
LOS	line-of-sight
LP	linking protection
LQA	link quality analysis
LSB	least-significant bit
LSU	link setup
MAC	media access control (protocol)
MB	megabyte(s)
MDL	multicast data link
MDR	medium data rate
MELP	multiple excitation linear prediction (vocoder)
MIL-STD	military standard (United States)
MLD	mid-latitude disturbed (channel condition)
MLSE	maximum-likelihood sequence estimator
MF	medium frequency
MIMO	multiple-input-multiple-output

MHz	megahertz (millions of cycles per second)
MSB	most-significant bit
NATO	North Atlantic Treaty Organization
NVIS	near-vertical incidence skywave
OFDM	orthogonal frequency division multiplexing
PAR	peak-to-average ratio
PDU	protocol data unit
PSK	phase-shift keying
PTM	point-to-multipoint (calling)
PTP	point-to-point (calling)
QAM	quadrature amplitude modulation
RF	radio frequency
RLSU	robust link setup
RX	receive
SAP	subnetwork access point (as defined in STANAG 5066)
SATCOM	satellite communications
SMTP	simple mail transfer protocol
SNDR	signal-to-noise-density ratio
SNR	signal-to-noise ratio
SPAWAR	Space and Naval Warfare Systems Command (United States)
SPNDR	signal-power-to-noise-density ratio
SSB	single sideband
SSN	smoothed sunspot number
STANAG	standardization agreement (NATO)
TAC	technical advisory committee (for standards development)
TCM	trellis-coded modulation
TCP	transmission control protocol
TDMA	time division multiple access
TGC	transmit gain control
TLC	transmit level control
TM	traffic management
TX	transmit
UAV	unmanned aerial vehicle
UDP	user datagram protocol
US	ultrashort (interleaver)
US	United States
VOACAP	Voice of America Coverage Analysis Program
WBALE	wideband automatic link establishment
WBHF	wideband HF

WID	waveform identification
WTRP	wireless token ring protocol
XOR	exclusive-or (logic operation)

About the Authors

Eric E. Johnson is a Professor Emeritus in the Klipsch School of Electrical and Computer Engineering at New Mexico State University (NMSU), and also leads a special projects group at the NMSU Physical Science Laboratory. He has published over 100 books, articles, papers, and technical reports in computer architecture and wireless technologies, including HF radio as well as mobile ad hoc and sensor networks. Research results from Dr. Johnson and his students have been incorporated into military standards in the US and NATO. He was the lead author of *Advanced High-Frequency Radio Communications*, published by Artech House in 1997.

Since the mid-1980s, Dr. Johnson has contributed to the development and standardization of HF radio technology. He currently chairs the NATO beyond-line-of-sight working group and the United States government/industry Technical Advisory Committee, which guides the development of United States Military Standards for HF radio.

His academic credentials include B.S. degrees in physics and electrical engineering, as well as an M.S. in electrical engineering from Washington University in St. Louis, and a Ph.D. in electrical and computer engineering from New Mexico State University.

Dr. Johnson is a registered professional engineer in New Mexico, holds a position as a chief scientist with Science Applications International Corporation (SAIC), and is also President of Johnson Research, a private consultancy. He served four years on active duty in the United States Army Signal Corps, and currently teaches short courses in HF radio for the Armed Forces Communications-Electronics Association (AFCEA), as well as specialized courses for other groups.

Eric Koski received his M.A. in philosophy from the University of Illinois at Urbana-Champaign in 1989 and his B.A. in general science (computer science) from the University of Rochester in 1982. He has worked for the Harris Corporation for 25 years; the primary focus of his work has been on HF radio communications technology, with side-excursions into tactical satellite and line-of-sight communications. He has research interests in the areas of wireless communications network and protocol design, software defined radio, and software product line engineering. He has authored or co-authored more than 20 published technical papers on topics in these areas, and has obtained four United States patents. He has been a key contributor to United States and international standards efforts, including those resulting in the HF communications standards MIL-STD-188-141B/C and NATO STANAG 4538: Automatic Radio Control System (ARCS).

William N. Furman received his B.S. and M.E. degrees in electrical engineering from Rensselaer Polytechnic Institute, Troy, NY, in 1982 and 1983, respectively. Since 1983, he has been employed by Harris Corporation, in both Melbourne, FL, and Rochester, NY, where he is currently a senior scientist and head of the Advanced Signal Processing Group. In 2001, he was recognized by Harris Corporation as a Harris Fellow for his sustained technical excellence and leadership in the field of advanced high-frequency radio communications.

His fields of interest are communications theory, forward error correction coding, digital signal processing, and the design of robust waveforms for use in challenging noise, interference, and dispersive channels. He has authored or co-authored over 30 papers on topics related to communications theory, coding, signal processing and networking and holds over 30 United States patents in these same areas.

He has worked in HF communications throughout his career at Harris and has been an implementer, designer, and key contributor to United States and NATO HF modem, automatic link establishment, and data link protocol standards.

Mark Jorgenson received his B.Sc. in electrical engineering from the University of Calgary in 1984. He spent three years as a combat systems engineering officer in the Canadian Navy before returning to school to receive his M.Sc. in electrical engineering at the University of Calgary in 1989. He worked as a research scientist at the Communications Research Center (CRC) in Ottawa, Canada. While at CRC, he was involved in research on modulation, coding, and receiver processing for HF data communications, and contributed to the development of several NATO STANAGs defining interoperable HF waveforms. He is one of the developers of the original 9600 bps QAM HF waveform and has worked with others in the community to define the MIL-STD and STANAG variants. He left CRC to found IP Unwired, a start-up that developed HF and V/UHF waveforms now in use by many navies around the world. IP Unwired was purchased by Rockwell Collins in 2006; he has continued to work with Rockwell Collins on interesting problems in HF and other bands.

John Nieto received his B.S. and M.S. degrees in electrical engineering from the University of Missouri-Rolla, in 1984 and 1985, respectively. Since 1985, he has been with the Harris Corporation, where he is currently a senior scientist in the Advanced Signal Processing Group of the Networking and Advanced Development Department of Harris RF Communications. His areas of interest are communications theory, waveform design, forward error correction coding, equalization, iterative demodulation, and simulation of digital communication systems. He has authored or co-authored over 50 papers on topics related to communications theory, coding, signal processing and networking and holds 47 United States patents in these same areas. He has worked in HF communications for the past 17 years and has been a key contributor to both U.S. and NATO HF waveform standards. In addition to this, he has been a key contributor to the signal processing algorithms used in a variety of Harris radio products covering the LF, HF, VHF, and UHF bands.

Index

W

Recent Titles in the Artech House Mobile Communications Series

John Walker, Series Editor

3G CDMA2000 Wireless System Engineering, Samuel C. Yang

3G Multimedia Network Services, Accounting, and User Profiles, Freddy Ghys, Marcel Mampaey, Michel Smouts, and Arto Vaaraniemi

802.11 WLANs and IP Networking: Security, QoS, and Mobility, Anand R. Prasad, Neeli R. Prasad

Achieving Interoperability in Critical IT and Communications Systems, Robert I. Desourdis, Peter J. Rosamilia, Christopher P. Jacobson, James E. Sinclair, and James R. McClure

Advances in 3G Enhanced Technologies for Wireless Communications, Jiangzhou Wang and Tung-Sang Ng, editors

Advances in Mobile Information Systems, John Walker, editor

Advances in Mobile Radio Access Networks, Y. Jay Guo

Applied Satellite Navigation Using GPS, GALILEO, and Augmentation Systems, Ramjee Prasad and Marina Ruggieri

Artificial Intelligence in Wireless Communications, Thomas W. Rondeau and Charles W. Bostian

Broadband Wireless Access and Local Network: Mobile WiMax and WiFi, Byeong Gi Lee and Sunghyun Choi

CDMA for Wireless Personal Communications, Ramjee Prasad

CDMA Mobile Radio Design, John B. Groe and Lawrence E. Larson

CDMA RF System Engineering, Samuel C. Yang

CDMA Systems Capacity Engineering, Kiseon Kim and Insoo Koo

CDMA Systems Engineering Handbook, Jhong S. Lee and Leonard E. Miller

Cell Planning for Wireless Communications, Manuel F. Cátedra and Jesús Pérez-Arriaga

Cellular Communications: Worldwide Market Development, Garry A. Garrard

Cellular Mobile Systems Engineering, Saleh Faruque

Cognitive Radio Techniques: Spectrum Sensing, Interference Mitigation, and Localization, Kandeepan Sithamparanathan and Andrea Giorgetti

The Complete Wireless Communications Professional: A Guide for Engineers and Managers, William Webb

EDGE for Mobile Internet, Emmanuel Seurre, Patrick Savelli, and Pierre-Jean Pietri

Emerging Public Safety Wireless Communication Systems, Robert I. Desourdis, Jr., et al.

The Future of Wireless Communications, William Webb

Geographic Information Systems Demystified, Stephen R. Galati

GPRS for Mobile Internet, Emmanuel Seurre, Patrick Savelli, and Pierre-Jean Pietri

GPRS: Gateway to Third Generation Mobile Networks, Gunnar Heine and Holger Sagkob

GSM and Personal Communications Handbook, Siegmund M. Redl, Matthias K. Weber, and Malcolm W. Oliphant

GSM Networks: Protocols, Terminology, and Implementation, Gunnar Heine

GSM System Engineering, Asha Mehrotra

Handbook of Land-Mobile Radio System Coverage, Garry C. Hess

Handbook of Mobile Radio Networks, Sami Tabbane

High-Speed Wireless ATM and LANs, Benny Bing

Interference Analysis and Reduction for Wireless Systems, Peter Stavroulakis

Introduction to 3G Mobile Communications, Second Edition, Juha Korhonen

Introduction to Communication Systems Simulation, Maurice Schiff

Introduction to Digital Professional Mobile Radio, Hans-Peter A. Ketterling

Introduction to GPS: The Global Positioning System, Ahmed El-Rabbany

An Introduction to GSM, Siegmund M. Redl, Matthias K. Weber, and Malcolm W. Oliphant

Introduction to Mobile Communications Engineering, José M. Hernando and F. Pérez-Fontán

Introduction to Radio Propagation for Fixed and Mobile Communications, John Doble

Introduction to Wireless Local Loop, Broadband and Narrowband, Systems, Second Edition, William Webb

IS-136 TDMA Technology, Economics, and Services, Lawrence Harte, Adrian Smith, and Charles A. Jacobs

Location Management and Routing in Mobile Wireless Networks, Amitava Mukherjee, Somprakash Bandyopadhyay, and Debashis Saha

LTE Air Interface Protocols, Mohammad T. Kawser

Mobile Data Communications Systems, Peter Wong and David Britland

Mobile IP Technology for M-Business, Mark Norris

Mobile Satellite Communications, Shingo Ohmori, Hiromitsu Wakana, and Seiichiro Kawase

Mobile Telecommunications Standards: GSM, UMTS, TETRA, and ERMES, Rudi Bekkers

Mobile Telecommunications: Standards, Regulation, and Applications, Rudi Bekkers and Jan Smits

Multiantenna Digital Radio Transmission, Massimiliano Martone

Multiantenna Wireless Communications Systems, Sergio Barbarossa

Multi-Gigabit Microwave and Millimeter-Wave Wireless Communications, Jonathan Wells

Multipath Phenomena in Cellular Networks, Nathan Blaunstein and Jørgen Bach Andersen

Multiuser Detection in CDMA Mobile Terminals, Piero Castoldi

OFDMA for Broadband Wireless Access, Slawomir Pietrzyk

Personal Wireless Communication with DECT and PWT, John Phillips and Gerard Mac Namee

Practical Wireless Data Modem Design, Jonathon Y. C. Cheah

Prime Codes with Applications to CDMA Optical and Wireless Networks, Guu-Chang Yang and Wing C. Kwong

Quantitative Analysis of Cognitive Radio and Network Performance, Preston Marshall

QoS in Integrated 3G Networks, Robert Lloyd-Evans

Radio Engineering for Wireless Communication and Sensor Applications, Antti V. Räisänen and Arto Lehto

Radio Propagation in Cellular Networks, Nathan Blaunstein

Radio Resource Management for Wireless Networks, Jens Zander and Seong-Lyun Kim

Radiowave Propagation and Antennas for Personal Communications, Third Edition, Kazimierz Siwiak and Yasaman Bahreini

RDS: The Radio Data System, Dietmar Kopitz and Bev Marks

Resource Allocation in Hierarchical Cellular Systems, Lauro Ortigoza-Guerrero and A. Hamid Aghvami

RF and Baseband Techniques for Software-Defined Radio, Peter B. Kenington

RF and Microwave Circuit Design for Wireless Communications, Lawrence E. Larson, editor

Sample Rate Conversion in Software Configurable Radios, Tim Hentschel

Signal Processing Applications in CDMA Communications, Hui Liu

Smart Antenna Engineering, Ahmed El Zooghby

Software Defined Radio for 3G, Paul Burns

Spread Spectrum CDMA Systems for Wireless Communications, Savo G. Glisic and Branka Vucetic

Technologies and Systems for Access and Transport Networks, Jan A. Audestad

Third-Generation and Wideband HF Radio Communications, Eric E. Johnson, Eric Koski, William N. Furman, Mark Jorgenson, and John Nieto

Third Generation Wireless Systems, Volume 1: Post-Shannon Signal Architectures, George M. Calhoun

Traffic Analysis and Design of Wireless IP Networks, Toni Janevski

Transmission Systems Design Handbook for Wireless Networks, Harvey Lehpamer

UMTS and Mobile Computing, Alexander Joseph Huber and Josef Franz Huber

Understanding Cellular Radio, William Webb

Understanding Digital PCS: The TDMA Standard, Cameron Kelly Coursey

Understanding GPS: Principles and Applications, Second Edtion, Elliott D. Kaplan and Christopher J. Hegarty, editors

Understanding WAP: Wireless Applications, Devices, and Services, Marcel van der Heijden and Marcus Taylor, editors

Universal Wireless Personal Communications, Ramjee Prasad

WCDMA: Towards IP Mobility and Mobile Internet, Tero Ojanperä and Ramjee Prasad, editors

Wireless Communications in Developing Countries: Cellular and Satellite Systems, Rachael E. Schwartz

Wireless Communications Evolution to 3G and Beyond, Saad Z. Asif

Wireless Intelligent Networking, Gerry Christensen, Paul G. Florack, and Robert Duncan

Wireless LAN Standards and Applications, Asunción Santamaría and Francisco J. López-Hernández, editors

Wireless Sensor and Ad Hoc Networks Under Diversified Network Scenarios, Subir Kumar Sarkar

Wireless Technician's Handbook, Second Edition, Andrew Miceli

For further information on these and other Artech House titles,including previously considered out-of-print books now available through our In-Print-Forever® (IPF®) program, contact:

Artech House	Artech House
685 Canton Street	16 Sussex Street
Norwood, MA 02062	London SW1V 4RW UK
Phone: 781-769-9750	Phone: +44 (0)20 7596-8750
Fax: 781-769-6334	Fax: +44 (0)20 7630-0166
e-mail: artech@artechhouse.com	e-mail: artech-uk@artechhouse.com

Find us on the World Wide Web at: www.artechhouse.com